Modeling Electrochemical Dynamics and Signaling Mechanisms in Excitable Cells with Pathological Case Studies

Modeling Electrochemical Dynamics and Signaling Mechanisms in Excitable Cells with Pathological Case Studies

TETSUYA WATANABE
President, Watanabe Institute of Mathematical Biology and
Watanabe Clinic of Oral Surgery, Japan

ELSEVIER

ACADEMIC PRESS
An imprint of Elsevier

Academic Press is an imprint of Elsevier
125 London Wall, London EC2Y 5AS, United Kingdom
525 B Street, Suite 1650, San Diego, CA 92101, United States
50 Hampshire Street, 5th Floor, Cambridge, MA 02139, United States
The Boulevard, Langford Lane, Kidlington, Oxford OX5 1GB, United Kingdom

Notices
Knowledge and best practice in this field are constantly changing. As new research and experience broaden our understanding, changes in research methods, professional practices, or medical treatment may become necessary.

Practitioners and researchers must always rely on their own experience and knowledge in evaluating and using any information, methods, compounds, or experiments described herein. In using such information or methods they should be mindful of their own safety and the safety of others, including parties for whom they have a professional responsibility.

To the fullest extent of the law, neither the Publisher nor the authors, contributors, or editors, assume any liability for any injury and/or damage to persons or property as a matter of products liability, negligence or otherwise, or from any use or operation of any methods, products, instructions, or ideas contained in the material herein.

Library of Congress Cataloging-in-Publication Data
A catalog record for this book is available from the Library of Congress

British Library Cataloguing-in-Publication Data
A catalogue record for this book is available from the British Library

ISBN 978-0-323-98803-2

For information on all Academic Press publications
visit our website at https://www.elsevier.com/books-and-journals

Publisher: Stacy Masucci
Acquisitions Editor: Elizabeth Brown
Editorial Project Manager: Ivy Dawn C. Torre
Production Project Manager: Sreejith Viswanathan
Cover Designer: Mark Rogers

Typeset by STRAIVE, India

Contents

Preface

Physical and mathematical bases for cell functioning are essential to a real understanding of the signals as we see them in cardiac and smooth muscles during electrophysiological study, and without the deep understanding, many measurable events such as generation and conduction of the impulse in neurons and muscles may remain unexplained. Now, we need to make a rigorous breakthrough by a physical approach to understand all the experimental information. Understanding of the electrochemical signaling mechanism is important to understand nerve and muscle disorders and will make an impact in the area of nerve and muscle behavior. Mathematical modeling and physical bases are introduced in this book for readers who are interested in these fields. The book will also cover the neuronal cell communication system in excitable cells, recognizing the most relevant mechanisms of cell communication. This book includes new findings in biotechnology being applied in medicine for biotechnologists and pathological case studies for clinicians as well as basic molecular physiology for physiologists and pharmacologists. Title "Modeling Electrochemical Dynamics and Signaling Mechanisms in Excitable Cells with Pathological Case Studies" is named since the electrochemical potential is oscillating in an exciting state of the living nerve and muscle cells. I hope this book will introduce a new field of physics, i.e., electrochemical dynamics and its application to living cells, and will give a good up-to-date insight into the field of electrophysiology. Background materials are covered in Chapters 2–4, basic mechanisms in Chapters 5–7, and medical applications are covered in Chapters 8–10. Introduction is added in Chapter 1, and I hope that will help both unfamiliar readers and readers without background in mathematical and physical interpretation.

This book was written after the "Physical Approach to Advanced Biology" and "Biophysical Physiology of the Cell, Lungs, and Excitable Tissues" published by Lambert Academic Publishing (2016) and "Biophysical Basis of Physiology and Calcium Signaling Mechanism in Cardiac and Smooth Muscle" by Academic Press (2018). Definitions and the concepts of background materials are more carefully interpreted. Taking a new angle and an approach will help one better understand the excitable tissues by providing a bridge between cell physiology and mathematical modeling.

Acknowledgments

It is my pleasure to thank one and all who have helped me publish this book. First, I thank Dr. Toshiomi Yoshida, professor emeritus of Osaka University, for offering his advice for this work. From 1972 to 1973, he visited the University of Pennsylvania for a research stay, and we have been close friends since then. I thank Dr. Akio Yoshida, visiting professor of Shizuoka University, for checking and offering helpful suggestions, despite being busy with other projects. We have been close friends studying physics and mathematics together since we both were high-school students. Although we went to different universities, he often visited me for studying physics and for swimming during the summertime because I lived near the beach. I became interested in physics under his influence. Professor Haruo Ito was a brilliant teacher who gave me an opportunity to do research on the effects of ACh and epinephrine on phosphorylase activities on rat heart when I was a student. He introduced me to Professor Domingo Aviado at Medical school, University of Pennsylvania, where I did research in pharmacology. Dr. Osamu Nishimura, professor emeritus of Doshisha University, was a graduate student of economics at the University of Pennsylvania. We became friends because we both love statistical mathematics. He helped me check Chapter 2. We meet annually and recall a good memory of life at the University of Pennsylvania.

Tetsuya Watanabe

CHAPTER 1
Introduction

1.1 Molecular interaction and increases in entropy associated with diffusion

In the isolated system at equilibrium, the system has the maximum entropy, and then the Boltzmann distribution appears. The microstates and entropy associated with molecular interaction increase every time molecules collide, exchange kinetic energy in the space, and finally stop at the maximum when equilibrium is attained in an isolated system. The number of locations molecules probably take in a space provides the number of locations where they exchange positions and relates to the number of probable redistributions of the kinetic energy, which determines the number of microstates on molecular interaction. Huge microstates and entropy of gas are mainly as a result of the molecular interaction associated with an exchange of their kinetic energy, occurring every time they collide in an isothermal space. The collision generates pressure, and the partial pressure of gaseous molecules is proportional to the concentration. An interacting mass of gas molecules has the potential to spread spontaneously if they can obtain a sparser space (see Section 2.5). We can see this phenomenon during gas exchange in the lungs (see Chapter 9).

Liquid molecules have the microstates and entropy that are mainly associated with molecular interaction like gas molecules but are different from gas molecules in that liquid molecules fill the space and bump continuously exchanging their energy and positions. If liquid is kept in a container where molecules exchange kinetic energy randomly every time they collide, molecules exchange positions but is not able to diffuse. Molecules in a container may have about the same average kinetic energy due to continuous bumping with others under the isothermal condition. Entropy is kept the same under constant temperature. Diffusion called current will occur if we give more space to spread. Liquid is different from gas in that the space is fully occupied by molecules, and the speed of diffusion is slower than that of gas. Bumping between molecules produces pressure which drives them to diffuse. Mixing of different kinds of molecules cause diffusion (see Section 2.6).

Modeling Electrochemical Dynamics and Signaling Mechanisms in Excitable Cells with Pathological Case Studies
https://doi.org/10.1016/B978-0-323-98803-2.00003-1

In a living cell, highly concentrated Ca^{2+} ions stored in the endoplasmic reticulum (ER) leak out through the ion channel, leading to electric potential change in the cell. If the ion conductivity of the channel is extremely low, ionic molecules may bump and travel zigzag through the narrow channel (see Section 5.9).

1.2 Functions of the cell membrane

The cell membranes are made of phospholipids, cholesterol, and proteins. The phospholipid bilayer forms the basic structure of the membrane. This self-orientation property of phospholipids leads the biological membrane into a closed spherical structure and reseals it by itself when torn out. Many of the transmembrane proteins that abut the extracellular fluid are glycoproteins. They not only give extra mechanical strength to the membrane but also make the cell capable of communicating with other cells. Some transmembrane proteins are channels and transporters. Only gases such as oxygen, carbon dioxide, and nitrogen, and small nonpolar lipid soluble molecules can diffuse freely through the phospholipid bilayer membrane along their concentration gradients. However, ions and polar molecules cannot diffuse through it, and their flux through the membrane requires channels or transporters that are made of columnar proteins firmly inserted into the phospholipid bilayer. Each kind of channels has a pore allowing the passage of only selected solute. Some channels are always open such as aquaporins (water channels) and potassium leak channels. But some channels have two states, open and closed, which can be controlled in response to the stimulus by binding a signaling molecule called a ligand to the receptor on the channel protein, or a change in membrane potential detected by means of the voltage sensor, or a mechanical force on the membrane by the pressure sensor. On the other hand, a transporter forms an intermediate complex with the solute, and a subsequent conformational change in the transporter causes translocation of the solutes to the other side of the membrane. Passive transport processes require no energy supply. They are designed for the diffusion of polar small molecules such as glucose and amino acids that are too large to go through membrane channels by themselves. Active transport processes are used for ions to diffuse across cell membranes against the electrochemical gradient by means of binding to a high-energy phosphate produced by ATP hydrolysis. Action of Na^+-K^+ ATPase is a good example of active transports.

The selective permeability of membrane creates osmosis, and interestingly along with the continuous actions of Na^+-K^+ ATPase, the membrane that is impermeable to Na^+ and permeable to K^+ at a resting state maintains the constant resting potential. Functions of living cells are regulated by charged or polar signaling small molecules called ligands. Some ligands can change the conformation of the ion channels in the cell membranes when they specifically bind to the receptors and open the channels for specific ions. The ion flux through the channels or transporters causes ionic current and can change the electric potential inside the cell (see Chapter 3). G protein-coupled receptors (GPCRs) are transmembrane proteins that detect ligands outside the cell and activate cellular responses via signal transduction pathways (see Section 7.3).

1.3 Oscillation of electrochemical potential in exciting living cells

We see diffusion when solids dissolve in liquid, different gases or liquids mix, solutes such as sodium ion spread in the fluid of the axon, or gases are dissolved in the blood. Diffusion is defined as the spread of solute moving from the area of higher concentration to the area of lower concentration along the concentration gradient. The concentrated area can be made by the injection of a solute into a solvent. Continuous bumping among different kinds of molecules produces pressure fluctuation that drives them down from the area of higher concentration to the area of lower concentration. The rate of diffusion depends on how much the molecules are concentrated in the certain area compared to its adjacent area, and it is inversely proportional to the resistivity of the tissue. The pressure produced by bumping between the same kind of molecule (A) drives them to push one away, while adjacent another kind of B molecule gets into the concentrated area of A molecules and replace A molecule with B molecule, thereby causing the diffusion by mixing. A higher concentration gradient suggests a higher probability of a collision between the same kind of molecules causing a replacement of one of the positions with another kind of adjacent molecule, thereby causing the diffusion by fast mixing. The speed of mixing is fast at first and gradually goes down. The diffusion by mixing will stop when its concentration becomes the same anywhere in the container, where entropy becomes maximized. Mixing makes entropy larger due to an increase in the number of probable incidences of redistribution of kinetic energy among different kinds of molecules under isothermal condition. We can see this phenomenon during gas exchange between blood and tissues (see Section 2.6.1).

If the solute is an ion or polar molecule, the speed of diffusion depends on the ion conductivity of the tissue and charges of the solute. Ions and polar molecules cannot pass through the phospholipid bilayer of the cell membrane and the scar tissue that have extremely low ion conductivity. They need channels and transporters to be able to pass through the membrane. If the ion resistivity of the channel is extremely high, the concentrated ionic molecules may bump and travel zigzag through it, which can make them oscillate. Diffusion of ions through a leak channel and injected ions in a model axon causes oscillations (see Sections 5.9 and 6.9), which may not be seen in sparse gaseous molecules in a large space. The oscillation of concentrated ions will cause the oscillation of electrical potential in living nerve and muscle fibers in an exciting state. By solving the cable equation, we found the membrane potential in the excited model axon shows a damped oscillation and withers over time and the distance that measures how far away from the origin of the current injection. This phenomenon is called diffusion and is evaluated mathematically. As the internal axoplasmic resistance is small, the oscillation damps fast. High friction is a prerequisite for consistent oscillation. Interestingly, similar damped oscillation occurs in the oscillatory release of stored Ca^{2+} in the sarcoplasmic reticulum (SR) through RyR2 channels (see Section 5.9). The cyclic release of stored Ca^{2+} from the SR or the endoplasmic reticulum (ER) was thought to be the basic source of the inherent automaticity. The RyR2 channels in the heart and IP3-gated Ca^{2+} channels in the intestine provide the places where oscillatory Ca^{2+} currents occur.

1.4 Physiological functions regulated by diffusion, attraction, and binding of ligands to their receptors

Physiological functions are thought to be results of the sum of diffusion, attraction, and binding of biological molecules in the cells. Different kinds of atoms have different abilities to attract electrons toward themselves, and the resultant bond has a partial ionic character. Electrostatic force attracts small polar molecules [1] and changes conformation of polar proteins. Cardiac and smooth muscles are activated by conformational change of proteins binding to a high-energy phosphate or to diffusible messengers, called ligands when binding to the polar site of a target proteins. A ligand is usually a small molecule which produces a signal by binding specifically to the site called a receptor on a target protein. Ligands may be hormones such epinephrin (EPI), or neurotransmitters such as GABA, glutamate, acetylcholine (Ach), and norepinephrine (NE), or substrates of enzyme, or second messengers such as cAMP, Ca^{2+}, inositol triphosphate (IP_3), and diacylglycerol (DG). Binding between a ligand and its receptor results from intermolecular forces such as electrostatic forces, which cause the conformational change of the target protein. The binding typically changes the three-dimensional shape of the target protein, which switch on various physiological effects depending on the target protein. The target protein may be (1) enzymes, (2) ligand-gated ion channels, or (3) G protein-coupled transmembrane proteins that bind selectively to hormones or the neurotransmitters. Neurons communicate one another by the neurotransmitter at the specialized small gap called a synapse. The neurotransmitter released from the terminals of presynaptic neuron diffuses across the synaptic cleft and activates the ligand-gated ion channels in the dendrites or cell body of the postsynaptic neuron to open, causing ion influxes to generate small localized-potentials called graded potentials in the dendrites and cell body. Graded potentials are either less negative than the resting membrane potential (>-70 mV), which makes the cell more excitable, or more negative compared to the resting potential (-70mV), which makes the cell not to excite. Influxes of positive ions such as Na^+, Ca^{2+} cause the excitatory graded potentials in the cell body. Efflux of positive ion like K^+ and influx of negative ion like Cl^- cause the inhibitory graded potentials in the cell body. Once formed, the graded potentials are decaying over distance. Graded potentials vary in size which are the results of summation of the excitatory and inhibitory postsynaptic potentials. When the sum of the postsynaptic potentials reaches the threshold (-55 mV) at axon hillock, it opens voltage-gated sodium channel in the axon. The sudden Na^+ influx initiates an action potential (see Chapter 6). G protein-coupled receptors detect ligands outside the cell and activate cellular responses by means of signal transduction pathways. The binding of the ligands to the receptors relays signals down to the interior of cardiac and smooth muscle cells and operates cellular metabolic responses via changing activities of the enzymes which catalyze production of second messengers such as cyclic AMP in cardiac and smooth muscles, and IP_3 and diacylglycerol in smooth muscles (see Section 4.4).

1.5 Basic mechanism of automaticity and its modulation

The cyclic release of stored Ca^{2+} from the SR produces the tone of the muscle and was thought to be the basic source of rhythmic cardiac automaticity.

Pale oval P cells are located centrally in the SA and AV nodes and so named as they are pale, primitive, and pacemaker cells. These cells are oval or rounded in contrast to the elongated shape of other myocardial cells. They are low-conductive cells with a shorter refractory period and have slower depolarization rate. The P cells in the SA node initiate rhythmical impulses automatically at a regular speed, that is, 60–80 beats per minute. Additionally, the P cells in the AV node are called latent pacemakers, which take over rhythmic impulse activities at the slightly slower pace, i.e., 40–60 beats per minute if the SA node stops initiating beats.

The sinoatrial (SA) and atrioventricular (AV) node of the heart have innervation of postganglionic parasympathetic and sympathetic nerves. Nerve impulses cannot initiate rhythmic contraction of cardiac muscle, but they can modulate the contraction rate and its intensity. The membranes of adjacent cardiac muscle fibers interlock each other. The impulse can pass directly from cell to cell through gap junctions. Thus, heart muscle fibers can do rhythmic coordinate contractions without the necessity of an individual innervation. However, its automaticity is under the control of the autonomic nerve innervation. The sympathetic nervous system mediates the excitability of the pacemaker cells by means of the second messenger of cAMP, which acts on HCN channels and activates cyclic AMP-dependent protein kinase (that phosphorylates the Ca^{2+} channels to open), and excite the pacemaker cells by another second messenger, i.e., Ca^{2+} by enhancing Ca^{2+} leak out of the SR in the heart. Cyclic AMP and Ca^{2+} by stimulation of sympathetic nerve is implicated in making cardiac muscles to be excitable (see Sections 7.4 and 7.9). On the other hand, activation of the IKACh channels begins with the release of ACh from the vagus nerve supplying the specialized myocardial tissue of the SA and AV nodes. ACh binds to the M_2 muscarinic acetylcholine receptors on pacemaker cells, which promotes the dissociation of the βγ-complex from α subunit of G_i protein [2] (see Section 7.3). Once the plug of IKACh channel is removed by binding to the βγ-complex, more K^+ ions flow out of the pacemaker cells causing hyperpolarization in a resting state during diastole [3]. In its hyperpolarized state, action potentials cannot be generated as quickly, which slows the heart rate (see Section 7.9).

The outstanding characteristic of the small intestinal muscle is its rhythmicity, which is alternate contraction and relaxation at a regular frequency. The basic source of pacemaker activity is thought to be rhythmic Ca^{2+} diffusion through IP_3-gated Ca^{2+} channels from the ER, which produces the tone of smooth muscle. However, the pacemaker activity can be modified by parasympathetic stimulation which causes an increase in cytosolic Ca^{2+} via opening of IP_3-gated Ca^{2+} channels and voltage-gated Ca^{2+} channels, thereby activating myosin ATPase and by sympathetic stimulation which causes a decrease in affinity between the light-chain kinase and Ca^{2+}-calmodulin, thereby inactivating myosin ATPase. A spike on each small rhythmic depolarization occurs when the circular muscle contracts. Each sharp spike may correspond to Na^{+} influx and the circular muscle contraction, and the small sinusoidal wave activity may relate to the oscillatory release of stored Ca^{2+} from the ER and correspond to the tone, but there is no indication of influx of extracellular Ca^{2+} during relaxation while the contractile stimulus such as parasympathetic stimulation and stretch are absent. ACh raised not only "the base membrane potential" that probably results from an increase in cytosolic Ca^{2+} but also "the sharp spike frequency," which results in large increases in the tonic tension and the rate of the circular muscle contractions. The raised base membrane potential probably due to high cytosolic Ca^{2+} is considered a result of opening of L-type Ca^{2+} channels by the activated protein kinase C (see Section 8.6). The second messenger in this case is triphosphate, diacylglycerol and Ca^{2+} instead of cyclic AMP and Ca^{2+} in the heart. The cyclic increase in cytosolic Ca^{2+} leads to the phosphorylation of the myosin light chain at serine, which induces the activation of myosin ATPase on the actin binding site. Energy released from the hydrolysis of ATP by the activated ATPase enables recocking of the myosin head to be ready for binding to actin and cyclic muscle contractions (see Section 8.5).

1.6 Medical applications

1.6.1 Adjustment of peripheral circulation

The widening of the vascular bed decreases resistance in segments of comparable length. Resistance vessels are small arteries and arterioles chiefly, and capillaries and smallest venules. In most peripheral part of the circulation, flow is controlled by opening of the resistance vessels so with the elevation of tissue metabolism flow increases, i.e., the vessels of digestive tracks dilate after eating while the vessels of brain dilate during study. Local cardiovascular control mechanisms control the radius of peripheral arteries and influence cardiac output. They have been shown to operate in the absence of central nervous system and peripheral autonomic innervation. The smallest blood vessels have an inherent automaticity, which can be modulated by stimulation of nerve fibers, circulating hormone such as epinephrine, or local electric stimulation. The mechanism of the inherent automaticity may be due to the release of rhythmic Ca^{2+} from the ER. All smooth muscles of blood vessels have innervations of sympathetic nerve, which control the muscle tone. Stimulation of β_2 receptor causes vasodilatation. The administration of β_2 adrenergic receptor agonists or α adrenergic receptor antagonists relaxes small blood vessels.

There is another cellular mechanism that leads to vasodilatation. Activation of M_3 receptors on vascular endothelial cells induces synthesis of nitric oxide. Endothelium-derived relaxing factor was later identified as the gaseous second messenger, nitric oxide (NO). Nitric oxide molecules are free radicals that diffuse freely across cell membranes, and induce the smooth muscle of blood vessels to relax by activating guanylate cyclase, which catalyzes the formation of highly reactive second intracellular messenger of cyclic GMP from GTP (guanosine triphosphate). Cyclic GMP can activate cyclic GMP-dependent protein kinases. Cyclic GMP-dependent protein kinases stimulate the opening of Ca^{2+}-activated K^+ channels, leading to hyperpolarization and relaxation of the muscle fibers and also blocks the activity of phospholipase C that catalyzes the formation of inositol triphosphate (IP_3) from membrane phospholipid, phosphatidylinositol 4,5 diphosphate (PIP_2) by hydrolysis. The decreased amount of hydrolyzed product IP_3 reduces the discharge of the stored calcium ions from the endoplasmic reticulum, leading to vasodilatation (see Chapter 8).

1.6.2 Respiratory diseases

In healthy lungs, ACh by both neuronal and nonneuronal sources acts through muscarinic receptors to regulate physiological functions. Muscarinic M_3 receptors control smooth muscle tone, mucus secretion, and inflammation. In chronic obstructive disease and asthma, cholinergic mechanisms contribute to increased bronchoconstriction and mucus secretion that limit airflow. Human bronchial smooth muscle receives little or no sympathetic innervation, but it contains large number of β_2 adrenergic receptors. The stimulation of β_2 adrenergic receptors leads to the activation of adenylyl cyclase, increase in cyclic AMP, and the activation of cyclic AMP-dependent protein kinase (PKA), which results in phosphorylation of myosin light chain kinase and decreased affinity of the light chain kinase to the Ca^{2+}-calmodulin, causing smooth muscle relaxation, which can be also seen in blood vessels. β_2 adrenergic receptor agonists also increase Ca^{2+}-activated K^+ currents in airway smooth muscle, leading to membrane hyperpolarization and relaxation. Since epinephrine (EPI) is 10 – 50 times more potent than norepinephrine (NE) on the β_2 receptors, circulating hormone, EPI may play an important role in the relaxation of bronchial smooth muscle during sympathetic stimulation. Short-acting β_2 adrenergic agonists such as albuterol (salbutamol) are used to increase airflow in conditions such as asthma and chronic obstructive pulmonary disease (COPD). In asthma, airflow limitation is generally fully reversible and is caused by bronchoconstriction. Muscarinic antagonists such as ipratropium bromide increase airflow in asthma by blocking cholinergic tone and by blocking reflex bronchoconstriction mediated by the vagus nerves. However, in COPD airflow limitation is not fully reversible, and is caused by structural changes and narrowing of peripheral airways along with parenchymal destruction. Muscarinic antagonists may increase airflow in COPD by blocking cholinergic tone at airway smooth muscle, but effects are limited.

Pneumonia is an inflammation of the lung affecting primarily the small air sacs known as alveoli. In pneumonia, the partial lung collapse occurs due to reduced distensibility of the lung as the air sacs fill with fluid or pus, and the fused basement membranes may be separated by interstitial exudate. The ventilation is impaired in the pathological states as oxygen is unable to reach bloodstream due to a decrease in the diffusion capacity. Typically, symptoms include some combination of sputum, cough, chest pain, fever, difficulty of breathing, and high respiratory rate. Pneumonia is an infection usually caused by viruses or bacteria. Factors predisposed to pneumonia include chronic obstructive pulmonary disease, asthma, diabetes, heart failure, a history of smoking, and cystic fibrosis. Diagnosis is often based on symptoms and physical examination. The patchy consolidations of the part of the lobe in chest X-ray may confirm the diagnosis with the help of blood tests and the examination of the sputum (see Chapters 8 and 9).

1.6.3 Cardiac arrhythmias

Action potentials of pacemaker cells and Purkinje fibers respond to extra stimuli differently. The ion conductivity of pacemaker cells is small. Pacemaker cells conduct impulses in a slow manner. The refractory period of the pacemaker cells is shorter, but magnitude of refractoriness depends on time. The ion conductivity of Purkinje cells is large. Purkinje cells conduct impulses in a fast manner. The refractory period of Purkinje fiber is longer, and magnitude of its refractoriness depends on the number of Na^+ channels recovered from inactivation. In the refractory period, impulse cannot travel in the cell at all. The degree of recovery of Na^+ channels from inactivation can be altered by Na^+ channel blocker such as local anesthetics, which causes a prolonged refractory period. The extra impulse may partially interfered with the preceding impulse from the SA node in a repolarization phase, depending on the degree of recovery of the sodium channels from inactivation. An ectopic action potential before completion of normal ventricular repolarization can be superimposed on the preceding pulse from the pacemaker cells in the SA node.

Arrhythmias occur if the disorganized impulse originating in the ectopic focus somewhere in the heart interferes with normal transmission of an impulse from the SA. The transmission of the impulse from the SA node will be disturbed if an ectopic focus in the heart fires a depolarization wave during or right after repolarization of the preceding impulse from the SA node. If an ectopic site is in the ventricles, the SA node that is out of refractory can fire off a pulse without any interference, and the atria can contract right on schedule, but the ventricles in the refractory period cannot contract following the atrial contraction. There arises a compensatory pause between ventricular contractions, bringing more ventricular filling for a strong contraction.

The cells in the AV node, bundle of His, and Purkinje fibers are called latent pacemakers. Each time the SA node fires, it resets all the successive slower ones not to fire until the pulse comes from the SA node. If the SA node stops, then the AV node would take over rhythmic impulse activities at a slightly slower pace. An ectopic atrial signal usually enters the SA node, depolarizes it, and resets its timing, causing a noncompensatory pause, which can be seen in ECG as changes in the PP interval that contain the ectopic beat, that is, the PP interval containing an ectopic beat in-between is less than two times of the normal PP interval.

If the atrial rate is extremely high like in the atrial fibrillation, most of the impulses cannot pass through the AV node to the ventricles due to refractory properties, and its conduction rate through the AV node can be reduced by the administration of Ca^{2+} channel blockers since the transmission between the slow-response cells depends on Ca^{2+} diffusion through gap junctional channels.

Low conductivity of the cells with high Ca^{2+} as seen in ischemic damage or under sympathetic stimulation, or low K^+ seen in hypokalemia may predispose latent pacemakers such as the AV node, the bundle of His, and Purkinje fibers, or cardiac muscle cells to get enhanced automaticity. Low conductivity and excitement of the cardiac cells may be prerequisite for the cells to obtain latent automaticity. High Ca^{2+} under sympathetic stimulation is thought to be a basis of the premature contractions since it not only reduces the ion conductivity of the tissue but also makes the muscle excitable.

The Ca^{2+} overload, valvular heart disease, or cardiomyopathy may predispose tachyarrhythmias to the ventricular fibrillation (see Chapter 10).

1.7 Conclusions

In liquid, driving force of diffusion is pressure produced by bumping molecules with kinetic energy. In a high concentrated spot of certain molecules, the pressure produced by bumping between them pushes the other kind of neighbor molecules away, while some of the other kind of molecules get into the concentrated area of certain molecules, thereby causing the diffusion by mixing. The concentrated area can be made by an injection of a solute into a solvent in a container. Diffusion of ions through a leak channel and injected ions in a model axon causes oscillations, which may not be seen in sparse gaseous molecules in a large space. Physiological functions are thought to be results of the sum of diffusion, attraction, and conformational changes of receptor proteins binding to biological molecules, i.e., ligands. Cardiac and smooth muscles are activated by conformational changes of proteins binding to a high-energy phosphate or diffusible Ca^{2+} binding to polar receptor proteins such as ion channel protein and calmodulin. The small sinusoidal wave activity in the muscle fiber would relate to the oscillatory release of stored Ca^{2+} from the ER or the SR and correspond to the muscle tone. In the refractory period, an impulse cannot travel in the cell at all as a depolarization of membrane potential cannot generate an action potential due to the inactivated voltage-gated Na^+ channels. The degree of recovery of Na^+ channels from inactivation can be altered by Na^+ channel blocker, which causes a prolonged refractory period. Arrhythmias occur if the disorganized impulse originating in the ectopic focus interferes with normal transmission of an impulse from the SA. The cell with high Ca^{2+} under sympathetic stimulation or ischemic condition is considered as a candidate for an ectopic focus of the premature contractions since it not only has low ion conductivity but also is excitable.

References

[1] T. Watanabe, Biophysical Basis of Physiology and Calcium Signaling Mechanism in Cardiac and Smooth Muscle, Academic Press, San Diego, CA, 2018 (Chapter 3).

[2] T.T. Ivanova-Nikolova, E.N. Nikolov, C. Hansen, J.D. Robishaw, Muscarinic K+ channel in the heart: modal regulation by G protein βγ subunits, J. Gen. Physiol. 112 (2) (1998) 199–210.

[3] L.G. Hommers, M.J. Lohse, M. Bünemann, Regulation of the inward rectifying properties of G-protein-activated inwardly rectifying K^+ (GIRK) channels by $G_{\beta\gamma}$ subunits, J. Biol. Chem. 278 (2) (2003) 1037–1043.

CHAPTER 2

Statistical dynamics and molecular interaction

2.1 Introduction to statistical thermodynamics

To know the number of probable energy distributions over the oscillators of molecules is essential when solving various chemical events mathematically. According to Boltzmann, energy is assumed to be countable. Boltzmann used the statistical concept to describe the definition of entropy.

The average entropy per oscillator of a solid or a molecule was defined statistically by Boltzmann as $S = k \ln \Omega$, where Ω is the number of probable energy distributions over the oscillators of solid molecules, and k is the Boltzmann constant. The well-known equation that entropy is proportional to the logarithm of the number of microstates was defined mathematically by introducing the number of probable energy distributions over an assembly of energy acceptors and justified if the definition is consistent with the classical definition [1]. The energy acceptor in the Boltzmann distribution model is assumed to be the simple oscillator which can hold any amount of energy, but the average energy of the oscillators is set to (kT). The oscillator is thought to be a pair of bonding electrons. Each oscillator can receive any amount of energy between zero and the total amount which depends on the temperature. The oscillator in a solid is independent and does not exchange energy or positions with others.

Light smaller molecules with a small intermolecular attraction tend to become liquid or gas at room temperature. Gaseous molecules obtain different amounts of kinetic energy since they can travel freely in a large space with less frequency of collisions. The deviation of translational kinetic energy of molecules is actually huge. Gases have a huge number of probable energy distributions and entropy that are chiefly associated with interaction called a collision in addition to molecular vibration and rotation. The kinetic energy of molecules is redistributed randomly at the location where every time collision occurs. Entropy of gas increases when the gas obtains more space. Liquid molecules have the microstates and entropy that are mainly associated with interaction like gas molecules but are different from gas molecules in that liquid molecules fill the space and bump continuously

Modeling Electrochemical Dynamics and Signaling Mechanisms in Excitable Cells with Pathological Case Studies
https://doi.org/10.1016/B978-0-323-98803-2.00010-9

exchanging their energy and positions. As a result of continuous interaction, liquid molecules may have about the same amount of average kinetic energy under isothermal equilibrium. An increase in entropy of liquids occurs when mixing starts.

The internal energy of a mass of molecules (U) is defined statistically as the expected value (E) of the net energy of particles (E). The internal energy can be the kinetic energy of microscopic motion of the system's particles from translations, rotations, and oscillations, and the potential energy associated with microscopic forces, including chemical bonds and the electrostatic attraction of molecules. The internal energy increases proportionally to temperature, and the increase can change a solid into liquid and gas. It cannot be measured directly but can be calculated in case of an ideal gas. The difference is determined by thermodynamic processes that carry the system between the reference state and the current state of interest.

$$U \equiv \mathrm{E}(E) = \sum_i E_i p_i = \sum_i \left(\varepsilon_{n_{1,i}} + \varepsilon_{n_{2,i}} + \varepsilon_{n_{3,i}} + \cdots\cdots + \varepsilon_{n_{N,i}}\right) p_i$$

where the system energy (E) is considered as the sum of amounts of energy ($\varepsilon_1, \varepsilon_2, \varepsilon_3, \varepsilon_4, \ldots, \varepsilon_i$) possessed by particles, n_1, n_2, n_3, $\ldots$, n_N.

If particles are identical, the internal energy is defined as the average amounts of energy particles possess times the number of particles in the system.

$$U \equiv \mathrm{E}(E) = N \cdot \mathrm{E}(\varepsilon)$$

Enthalpy is defined as $H \equiv U + pV$. It depends on the state of the system. Like internal energy, it has a definite value when the state of the system is specified by two of the state properties such as p, V, or T. It does not depend on the path of a process, that is, enthalpy is a state function.

If the system changes under constant external pressure, p_{ext},

$$\Delta H = \Delta U + p_{\mathrm{ext}} \Delta V = \left(q_p - p_{\mathrm{ext}} \Delta V\right) + p_{\mathrm{ext}} \Delta V$$

$$\Delta H = q_p$$

The heat transfer at constant pressure (q_p) between the system and the surroundings is equal to the change in enthalpy, and the heat withdrawn from the surroundings is defined as positive. In an *endothermic* process, ΔH is positive. In an *exothermic* process, ΔH is negative.

The relationship between the internal energy and the enthalpy under constant pressure (p) is as follows:

$$\Delta U = q_p - p_{\mathrm{ext}} \Delta V = \Delta H - p_{\mathrm{ext}} \Delta V, \quad \Delta H = \Delta U + p_{\mathrm{ext}} \Delta V$$

For gases, ΔH and ΔU differ by 1%–2% due to volume change (ΔV) during the process. For liquids and solids, the difference between ΔU and ΔH is negligible. $\Delta H = q_p$ is measurable. If $q_p = 0$ and $\Delta V \approx 0$, $\Delta H = \Delta U \approx 0$.

2.2 Energy distribution in a solid

The energy acceptor in the Boltzmann distribution model is assumed to be the simple oscillator binding atoms in a solid which can hold any amount of energy. Energy is randomly distributed to the oscillators, each of which has an intrinsic frequency. The number of oscillators depends on the number of atoms in a solid. The larger size of a solid has more atoms and more oscillators since each atom of a solid has three oscillators at most. Generally, a large polyatomic solid has more oscillators which can hold more energy than a small solid at the same temperature. That is why the average heat capacity of a large polyatomic solid is usually larger than that of a small one at the same temperature. According to Boltzmann, energy was assumed to be countable, and each oscillator receives any number of energy bits (i) between zero and the total amounts of energy (E), and the oscillators are independent and do not exchange energy or positions among them. Both the total amounts of energy and the number of oscillators (N) in a solid are assumed to be constant in an isolated system.

The number of permutations of N distinguishable objects is $N!$. This is easy to explain. Any of the N objects can be in the first position, any of the other ($N-1$) in the second position, any of the remaining ($N-2$) in the third position, and so on. If the N objects are indistinguishable, the number of permutations is $N!/N! = 1$. If N is the sum of two different objects, N_A and N_B, the number of permutations is $N!/N_A!\, N_B!$.

Question 2.1 How many probable energy distributions are there if we put a total of six energy bits into four Boltzmann oscillators in a solid? The oscillator can hold any number of the energy bits between zero and six. The oscillators cannot exchange positions or energy among them.

Answer First, we assume that six identical bits of energy are distinguishable as well as oscillators. Then six identical bits of energy are numbered from one to six. Each oscillator has any amount of energy bits. We should admit special cases where some molecules have no energy bits, and one oscillator has all of the six energy bits for itself. We also admit some oscillators have the same number of energy bits.

We may write the contents of the molecules in order as follows. For example, for four molecules ($N=4$) with a total energy of six ($E=6$), we may have (, 1 2 3, 4 5, 6) which means none in the first oscillator; 1, 2, 3 in the second; 4 and 5 in the third; and 6 in the fourth; or (, 5 2 4, 3, 6 1), which means none in the first oscillator; 5, 2, and 4 in the second; 3 in the third; and 6 and 1 in the fourth; or (2, 5 4, 6 1, 3), which means 2 in the first oscillator; 5 and 4 in the second; 6 and 1 in the third; and 3 in the fourth. The first and second permutations show the same number of energy bits in the first and second oscillators. The first and third permutations show the same number of energy bits in the fourth oscillator.

Now we replace the commas representing boundaries between distinguishable oscillators with letters, which can be in any order, although the first set of numbers still denotes the contents of the first oscillator, the second set denotes the contents of the second, and so on. The first permutation above can be written as (A 1 2 3 B 4 5 C 6), (A 1 2 3 C 4 5 B 6), (B 123A45C 6), (B 1 2 3 C 4 5 A 6), (C 1 2 3 A 4 5 B 6), or (C 1 2 3 B 4 5 A 6).

Clearly, in any case, the total number of permutations, which is the number of probable ways of arranging six numbers and three letters, is

$$\Omega = (E + N - 1)! = (6 + 3)! = 9!$$

The 9! different outcomes are called microstates in statistical mechanics. The total number of microstates is 9!. *The number of microstates is the number of probable energy distributions over oscillators in a solid.*

Let us count the number of ways of putting a total of E energy bits into N oscillators. The oscillator cannot exchange position and energy. Because the number of boundaries is fewer than the number of oscillators by one, we have $(N-1)$ boundaries instead of N. There is a lot of overcounting in the way of counting permutations because we cannot distinguish bits of energy from others and cannot also distinguish boundaries from others since oscillators are indistinguishable. To correct the counting of permutations, we must divide by the number of ways of writing the energy numbers, which is $E!$, and by the number of ways of arranging the boundary labels, which is $(N-1)!$. After correction is made, we can obtain the number of permutations, Ω_{all} when distributing the total of E indistinguishable energy bits over the N oscillators. We call Ω_{all} the total number of microstates.

$$\Omega_{\text{all}} = \frac{(E + N - 1)!}{E!(N - 1)!} \tag{2.1}$$

Ω_{all} increases as E and N increase.

A large polyatomic molecule possesses many oscillators, and therefore it has a larger entropy at the same temperature according to Eq. (2.1). It can hold more energy and has a larger heat capacity than a small molecule.

2.3 Energy exchange among interacting solids

We have been studying about the number of probable energy distributions over oscillators in a solid. Now, we consider interacting solids which are exchanging their energy among them. We will discuss about the number of probable energy distributions over oscillators in the interacting solids in an isolated container. The solids are bumping and exchanging their energy randomly. Every time they bump, they will redistribute their energy over their oscillators. Then the energy distribution over their oscillators will change every time they bump.

Our simple model is the system of two identical interacting solid molecules in contact in a container every time they bump, and each solid molecule has the same number of oscillators. The solid molecules A and B with energy E_A and E_B, respectively, are exchanging energy. The total energy (E_{tot}) of two solid molecules is supposed to be constant since the system is isolated (Fig. 2.1).

$$E_{tot} = E_A + E_B$$

There are multiple values of $\Omega_{all}(E_A, E_B) = \Omega_{all}(E_A)\Omega_{all}(E_B)$ constrained by the temporary values of E_A and E_B in the isolated interacting system. $\Omega_{all}(E_A, E_B)$ is the function of E_A and E_B. Now, the total energy of E_A and E_B is set to 8 and is

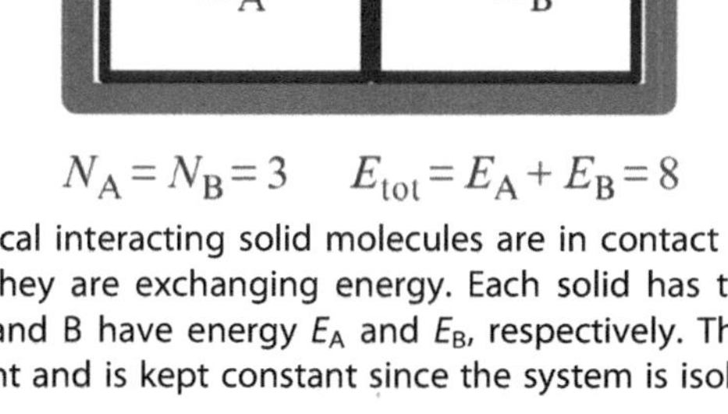

Fig. 2.1 Two identical interacting solid molecules are in contact in a container every time they bump. They are exchanging energy. Each solid has three oscillators. The solid molecules A and B have energy E_A and E_B, respectively. The total energy (E_{tot}) of two solids is eight and is kept constant since the system is isolated.

kept constant. Each solid has three oscillators. Two solids with E_A and E_B are exchanging energy. The number of microstates corresponding to E_A and E_B and the total microstates, $\Omega_{all} = \Omega(E_A)\,\Omega(E_B)$ are calculated using the following equation as shown in Table 2.1.

$$\Omega_{all}(E_A, E_B) = \Omega_{all}(E_A) \cdot \Omega_{all}(E_B) = \frac{(E_A + N_A - 1)!}{E_A!(N_A - 1)!} \cdot \frac{(E_B + N_B - 1)!}{E_B!(N_B - 1)!}$$

where N denotes the number of oscillators, and $E_A + E_B$ is constant.

If $E_{tot} = 8$ and $N_A = N_B = 3$, the number of overall microstates possibly appearing in the system over a long-time scale is 1287 as shown in Table 2.1. Each of the 1287 microstates, that is, each of the 1287 probable energy distributions over oscillators in the interacting solids, is equally probable. If you look at the system at any instant, you are equally likely to find one microstate in any of the 1287 microstates. This assumption is called the fundamental assumption of statistical mechanics [2].

In Table 2.1, you will see the maximum of $\Omega_{all}(E)$ at $E_A = E_B = E_{tot}/2 = \langle E_{tot} \rangle = 4$. In equilibrium, E_A becomes about the same as E_B. So the maximum number of Ω_{all} appears at the average energy, $\langle E_{tot} \rangle$ when equilibrium is reached.

Energy is redistributed randomly over oscillators every time a collision occurs and the distribution changes. Now, we will make a mathematical justification of the number of probable energy distributions over numerous oscillators in the interacting solids near equilibrium in the isolated container where solids are in contact and exchange energy every time they bump. Each polyatomic solid has a huge number of oscillators (N) that bond atoms,

Table 2.1 Two identical interacting solids in contact and isolated from the surroundings.

E_A	$\Omega(E_A)$	E_B	$\Omega(E_B)$	$\Omega(E_A)\,\Omega(E_B)$
0	1	8	45	45
1	3	7	36	108
2	6	6	28	168
3	10	5	21	210
4	15	4	15	225
5	21	3	10	210
6	28	2	6	168
7	36	1	3	108
8	45	0	1	45
				Total = 1287

and the total energy (E) in the isolated container is assumed to be constant. The peak of Ω_{all} will become higher when N gets larger, which will be justified mathematically as follows.

If N is huge, Eq. (2.1) becomes $\Omega_{all}(E) = \frac{(E+N-1)!}{E!(N-1)!} \approx \frac{(E+N)!}{E!N!}$.

We will take the natural logarithm of Eq. (2.1), apply Stirling's approximation, and rewrite it as

$$\begin{aligned} \ln\Omega_{all}(E) &\approx \ln\frac{(E+N)!}{E!N!} = \ln(E+N)! - \ln(E!) - \ln(N!) \\ &\approx (E+N)\ln(E+N) - (E+N) - E\cdot\ln E + E - N\cdot\ln N + N \\ \ln\Omega_{all}(E) &\approx (E+N)\ln(E+N) - E\cdot\ln E - N\cdot\ln N \\ \ln(E+N) &= \ln E\left(1+\frac{N}{E}\right) = \ln E + \ln\left(1+\frac{N}{E}\right) \end{aligned} \tag{2.2}$$

Since we assume $E \gg N$, N/E is extremely small and so we obtain the following approximation using Taylor's expansion of the logarithm.

$$\ln(1+x) \approx x \quad \text{as}\, x \ll 1$$

$$\ln(E+N) \approx \ln E + \frac{N}{E}$$

Substituting $\ln E + \frac{N}{E}$ for $\ln(E+N)$ in Eq. (2.2), we obtain

$$\begin{aligned} \ln\Omega_{all}(E) &\approx (E+N)\left(\ln E + \frac{N}{E}\right) - E\cdot\ln E - N\cdot\ln N \\ &= N\cdot\ln E - N\cdot\ln N + N + \frac{N^2}{E} \\ \ln\Omega_{all}(E) &\approx N\cdot\ln\left(\frac{E}{N}\right) + N + \frac{N^2}{E} \end{aligned} \tag{2.3}$$

The last term is negligible compared to the rest with the assumption of $E \gg N$.

$$\begin{aligned} \ln\Omega_{all}(E) &\approx N\left(\frac{\ln E}{N}\right) + N \\ \Omega_{all}(E) &\approx e^{N\left(\frac{\ln E}{N}\right)} e^N = \left(\frac{E}{N}\right)^N e^N = \left(\frac{eE}{N}\right)^N \end{aligned} \tag{2.4}$$

It is easily understandable that Ω is an extremely large number and will increase a lot by a small change of E according to Eq. (2.4).

Let us apply the calculated result to the system with two identical interacting solids model.

The solids A and B have the same number of oscillators and different amount of energy E_A and E_B, respectively. The total system energy, E_{tot} is the sum of E_A and E_B, and it is constant.

$$\Omega_{all}(E_A, E_B) \approx \left(\frac{eE_A}{N}\right)^N \left(\frac{eE_B}{N}\right)^N = \left(\frac{e}{N}\right)^{2N} (E_A E_B)^N \tag{2.5}$$

In equilibrium, the number of total microstates, $\Omega_{all}(E_A, E_B)$ reaches the peak at $E_A = E_B = E_{tot}/2 = \langle E \rangle$.

$$\Omega_{max} \approx \left(\frac{e}{N}\right)^{2N} \left(\frac{E_{tot}}{2}\right)^{2N} \tag{2.6}$$

In order to find how the graph of $Y = \Omega_{all}(E_A, E_B)$ looks like near equilibrium, we substitute $\frac{E_{tot}}{2} + \Delta E$ and $\frac{E_{tot}}{2} - \Delta E$ for E_A and E_B, respectively, in Eq. (2.5). Ω_{all} is the function of E_A and E_B.

$$\Omega_{all}(E_A, E_B) \approx \left(\frac{e}{N}\right)^{2N} (E_A E_B)^N = \left(\frac{e}{N}\right)^{2N} \left[\left(\frac{E_{tot}}{2}\right)^2 - \Delta E^2\right]^N \tag{2.7}$$

We take the logarithm to simplify the second factor of Eq. (2.7).

$$\begin{aligned}
\ln\left[\left(\frac{E_{tot}}{2}\right)^2 - \Delta E^2\right]^N &= N \cdot \ln\left[\left(\frac{E_{tot}}{2}\right)^2 - \Delta E^2\right] \\
&= N \cdot \ln\left[\left(\frac{E_{tot}}{2}\right)^2 \left(1 - \left(\frac{2\Delta E}{E_{tot}}\right)^2\right)\right] \\
&= N\left[\ln\left(\frac{E_{tot}}{2}\right)^2 + \ln\left(1 - \left(\frac{2\Delta E}{E_{tot}}\right)^2\right)\right] \\
&\approx N\left[\ln\left(\frac{E_{tot}}{2}\right)^2 - \left(\frac{2\Delta E}{E_{tot}}\right)^2\right]
\end{aligned}$$

By expressing the former equation exponentially, we obtain

$$\left[\left(\frac{E_{tot}}{2}\right)^2 - \Delta E^2\right]^N \approx e^{N\left[\ln\left(\frac{E_{tot}}{2}\right)^2 - \left(\frac{2\Delta E}{E_{tot}}\right)^2\right]}$$

Putting this back into Eq. (2.7), the distribution of $\Omega_{all}(E_A, E_B)$ near equilibrium follows the Gaussian distribution.

$$\Omega_{all}(E_A, E_B) \approx \left(\frac{e}{N}\right)^{2N}\left[\left(\frac{E_{tot}}{2}\right)^2 - \Delta E^2\right]^N$$
$$= \left(\frac{e}{N}\right)^{2N} e^{N \ln\left(\frac{E_{tot}}{2}\right)^2} e^{-N\left(\frac{2\Delta E}{E_{tot}}\right)^2} = \Omega_{max}\, e^{-N\left(\frac{2\Delta E}{E_{tot}}\right)^2} \qquad (2.8)$$

As the number of total microstates falls to $1/e$ of its maximum value at $\Delta E = \frac{E_{tot}}{2\sqrt{N}}$, the width of the peak is still very narrow, and the width is $\frac{E_{tot}}{\sqrt{N}}$ as the number of microstates falls to Ω_{max}/e. If $N = 10^{20}$, the width is $E_{tot} \times 10^{-10}$. In the case of two large interacting solids, the maximum number of microstates, $\Omega_{max}(E_A, E_B)$ appears at $E_A = E_B = \langle E \rangle$ in equilibrium in an isolated system (Fig. 2.2). The results can be applied to more than two identical interacting solids in the isolated system. In an isolated container, each interacting solid will have about the same average energy near equilibrium when the probability of finding an oscillator with a certain energy in a solid follows the Boltzmann distribution (see Example 2.1 in Section 2.4), and the distribution of the total microstates called system microstates near equilibrium follows the Gaussian distribution. In equilibrium, the number of total microstates and entropy becomes the maximum in the isolated system and gets smaller as the equilibrium is broken. The result can be applied to liquid molecules which occupy a space, bump each other, and exchange energy and positions continuously. Interacting molecules would have about the same amounts of energy at equilibrium while bumping continues, then the number of total microstates becomes the maximum at equilibrium in the isolated container.

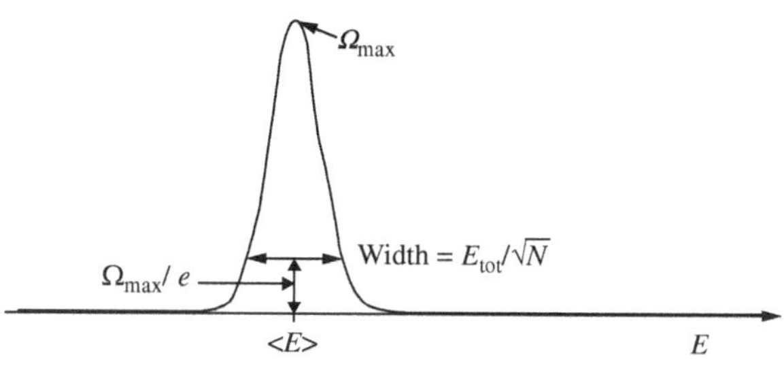

Fig. 2.2 The distribution of the total microstates near equilibrium in two large interacting solids. The width of the peak is very narrow, and the width is $\frac{E_{tot}}{\sqrt{N}}$ as the number of microstates falls to Ω_{max}/e. $\Omega_{max}(E) \approx \left(\frac{e}{N}\right)^{2N}\left(\frac{E_{tot}}{2}\right)^{2N}$ appears at $E_A = E_B = \langle E \rangle$ in equilibrium.

Question 2.2 Prove $e^{N\left(\frac{\ln E}{N}\right)}e^{N}=\left(\frac{eE}{N}\right)^{N}$.

Answer

If $X=N\left(\frac{\ln E}{N}\right)=\ln\left(\frac{E}{N}\right)^{N}$, then $e^{X}=\left(\frac{E}{N}\right)^{N}$.

$$e^{N\left(\frac{\ln E}{N}\right)}e^{N}=\left(\frac{E}{N}\right)^{N}e^{N}=\left(\frac{eE}{N}\right)^{N}$$

2.4 Statistical definition of entropy and the Boltzmann distribution to oscillators in a solid

A solid in an isolated system contains many oscillators (N). We can count the number of probable energy distributions over the energy acceptors if we know $\sum_{i=0}n_i=N$ and $E_{\text{total}}=\sum_{i=0}n_i\varepsilon_i$. Boltzmann assumed that the energy acceptor called the "oscillator" can hold any discrete number of energy bits from zero to the total amount of energy, but oscillators cannot exchange energy and positions. The average energy of the Boltzmann oscillators is set to kT indicating that the average energy is proportional to temperature. The number of oscillators holding various energy is denoted by n_0, n_1, n_2, ..., where n_0 is the number of oscillators with amounts of energy "0" (no energy), n_1 is the number of oscillators with amounts of energy "ε_1," n_2 is the number of oscillators with amounts of energy "ε_2," and so on. We will count the number of ways of distributing n_0 "0"s to the N oscillators.

$N\times(N-1)\times(N-2)\times\ldots\times(N-n_0+1)/n_0!$ permutations are obtained when putting n_0 "0"s into N oscillators. Since each energy bit is indistinguishable, we must divide it by n_0!

Successively n_1 "ε_1"s are put into the $(N-n_0)$ oscillators, n_2 "ε_2"s into the $(N-n_0-n_1)$ oscillators, and so on. The number of energy states increases every time an energy bit is put into a vacant oscillator.

The number of probable energy distributions over the isolated system containing N oscillators, Ω_N is calculated below using Stirling's approximation, $N!\approx e^{N\ln N}e^{-N}$ since N is assumed to be extremely large [1]. As N is extremely large, n_i can be considered as continuous variables.

$$\begin{aligned}\Omega_N&=\frac{N!}{n_0!n_1!n_2!n_3!\ldots\ldots}\\&\approx\frac{N^Ne^{-N}}{{n_0}^{n_0}\times{n_1}^{n_1}\times{n_2}^{n_2}\times{n_3}^{n_3}\times\cdots\times e^{-n_0-n_1-n_2-n_3\ldots}}\\&=\frac{N^N}{{n_0}^{n_0}\times{n_1}^{n_1}\times{n_2}^{n_2}\times{n_3}^{n_3}\times\cdots}\end{aligned}$$

where $\sum_{i=0} n_i = N$.

Let us rewrite in natural logarithmic form, and $p_i = n_i/N$ (probability for n_i to occur).

$$\begin{aligned}\ln \Omega_N &\approx N \cdot \ln N - (n_0 \ln n_0 + n_1 \ln n_1 + n_2 \ln n_2 + n_3 \ln n_3 \cdots) \\ &= N \cdot \ln N - \sum_{i=0} n_i \cdot \ln(n_i) = N \cdot \ln N - \sum_i p_i N \cdot \ln(p_i N) \\ &= N \cdot \ln N - \sum_i p_i N(\ln p_i + \ln N) \\ &= N \cdot \ln N - \sum_i p_i N \cdot \ln p_i - \sum_i p_i \cdot N \cdot \ln N = -\sum_i p_i N \cdot \ln p_i\end{aligned}$$

$$\ln \Omega_N = N \sum_{i=0} p_i \cdot \ln \frac{1}{p_i} \tag{2.9}$$

The number of probable energy distributions, Ω_N is the number of ways of distributing energy bits to its N oscillators in a solid. Eq. (2.9) is consistent with the fundamental assumption of statistical mechanics. If you look at the system at any instant, you are equally likely to find one microstate in any of all Ω microstates. If we have Ω microstates, the probability of finding any one microstate is $1/\Omega$, then the expected value of $\ln \frac{1}{p}$, i.e., $\sum_{i=0} p_i \cdot \ln \frac{1}{p_i}$ is expressed as $\ln\Omega$.

As in the interacting system discussed in Section 2.3, the total microstates are $\Omega_{\text{total}} = \Omega_A \Omega_B$.

Because entropy is a state function, it should be additive. $S_{\text{total}} = S_A + S_B$.

$$f(\Omega_A \Omega_B) = f(\Omega_A) + f(\Omega_B)$$

The only function that satisfies the above equation is $f = \ln \Omega$.

Then entropy was defined and described statistically as a function of $\ln \Omega$ by Boltzmann.

$$S \propto \ln \Omega$$

Using the Boltzmann constant k, the average entropy per oscillator in an isolated system containing many oscillators is expressed as $S = k \ln \Omega$ and it was discussed whether this statistical definition is consistent with the classical definition [1].

$$S = k \ln \Omega = \frac{S_N}{N} = \frac{k \ln \Omega_N}{N} = k \sum_{i=0} p_i \cdot \ln \frac{1}{p_i}$$

where $\Omega^N = \Omega_N$.

The above equation says that the average entropy of the oscillators is proportional to the logarithm of total microstates per oscillator that is the expected value of $\ln \frac{1}{p}$.

The maximizing entropy of many oscillators in an isolated system is the same as maximizing the microstates constrained by $\sum_{i=0} n_i = N$ and $E_{\text{total}} = \sum_{i=0} n_i \varepsilon_i$ which are constant. The Boltzmann distribution appears as the outcome of the maximizing entropy in an isolated system (see Example 2.1).

Example 2.1 In an isolated system, entropy is added every time an energy bit is distributed to an oscillator in a solid and finally becomes the maximum when equilibrium is obtained. The number of Boltzmann oscillators (N) and the total energy (E) are assumed to be constant. It is exciting to examine the Boltzmann distribution that appears when maximizing entropy by means of maximizing Ω_N.

$$\begin{aligned}\ln \Omega_N &= \ln\left(\frac{N!}{n_0! n_1! n_2! n_3! \cdots}\right)\\ &= N \ln N - N - (n_0 \ln n_0 - n_0 + n_1 \ln n_1 - n_1 + n_2 \ln n_2 - n_2 + n_3 \ln n_3 - n_3 \cdots)\\ &= N \ln N - \sum_i n_i \cdot \ln n_i\end{aligned}$$

If N and E are very large, n_i and ε_i are considered as continuous variables. When Ω_N is maximized,

$$d(\ln \Omega_N) = -\left(\sum_i \ln n_i + \sum_i n_i \cdot \frac{1}{n_i}\right) dn_i = 0 \tag{a}$$

Since $N = \sum_i n_i =$ constant, and $E_{\text{total}} = \sum_{i=0} n_i \varepsilon_i =$ constant in an isolated system,

$$(\alpha - 1)dN = (\alpha - 1)\sum_i dn_i = 0 \tag{b}$$

$$\beta \cdot dE = \beta \sum_{i=0} \varepsilon_i dn_i = 0 \tag{c}$$

We sum up the above Eqs. (a), (b), and (c) when Ω_N is maximized.

$$\sum_i (\ln n_i + \alpha + \beta \varepsilon_i) dn_i = 0$$

$$\ln n_i + \alpha + \beta \varepsilon_i = 0$$

$$n_i = e^{-\alpha - \beta \varepsilon_i} = n_0\, e^{-\beta \varepsilon_i}$$

where $n_0 = e^{-\alpha}$ is the number of oscillators which has no energy.

$$p_i = n_i/N = \frac{n_0}{N} \cdot e^{-\beta \varepsilon_i} \tag{2.10}$$

Maximizing entropy of the oscillators is the same as maximizing the number of microstates. Entropy is added every time an energy bit is distributed to an oscillator in a solid and finally becomes the maximum. When an isolated system gets the maximum entropy, the Boltzmann distribution Eq. (2.10) appears. According to the distribution, the probability of finding an oscillator with no energy is the maximum, and the probability of finding an oscillator with the total energy (E) is almost zero.

2.5 Interacting gaseous molecules in a space

Molecules with a small molecular weight tend to be gas at room temperature when their intermolecular attraction is extremely small. Gaseous molecules possess different amounts of kinetic energy since they can travel freely in a large space with less frequency of collisions. The deviation of translational kinetic energy of molecules is actually huge, and the velocity distribution which measures the probability of finding a molecule with a certain velocity in the unit range follows the Gaussian distribution [3]. Gases have the microstates and entropy that are associated with molecular vibration, rotation, and chiefly interaction called a collision. The collision generates pressure, and the pressure is proportinal to the concentration, i.e., the number of molecules per space. Molecules move from a higher concentration to a lower concentration according to the gradient. The kinetic energy of molecules is redistributed randomly between the molecules at a location every time collision occurs. Huge microstates and entropy are mainly as a result of the molecular interaction associated with an exchange of their kinetic energy occurring every time they collide in an isothermal space. *The number of locations molecules probably take in a space provides the number of locations where they exchange positions and relates to the number of probable redistributions of the kinetic energy, which determines the number of microstates on molecular interaction.* Huge microstates and entropy associated with the interaction depend on the number of molecules and a volume of the space they occupy since they determine the number of probable incidences of kinetic energy redistribution in the space if temperature is kept the same. The number of microstates associated with the interaction can be calculated statistically using the lattice model. That is how many probable locations molecules can take in the space.

Interacting gas molecules have the potential to diffuse spontaneously until the pressure becomes the same everywhere in the space if the molecules can obtain a lower pressure space.

Suppose we have a gaseous molecule whose molecular size is 10^{-30} m^3 and the total volume of the space is $1\,m^3$. A gaseous identical molecule may take 10^{30} probable locations. One molecule can enter one of the 10^{30} lattice sites in a space of $1\,m^3$. When gaseous particles collide in a space, they exchange their kinetic energy. The total number of probable energy redistributions can be calculated statistically by counting the number of locations molecules probably take in the space. The first one can take 10^{30} different positions where the energy redistribution can occur, the second can take $(10^{30}-1)$ different positions, the third can take $(10^{30}-2)$ positions, and so on. Let us calculate the number of microstates arising from the interacting N gaseous molecules in $1\,m^3$ under isothermal conditions.

The number of microstates for distinguishable N molecules is

$$\Omega_{trans,N} = 10^{30}\left(10^{30}-1\right)\left(10^{30}-2\right)\cdots\left(10^{30}-N+1\right) \approx 10^{30N}$$

For indistinguishable N molecules, $\Omega_{trans,N} \approx \frac{10^{30N}}{N!}$.

Since N is large, entropy of N molecules on translation can be calculated by applying Stirling's approximation, that is, $N! \approx e^{-N}N^N$. The entropy per $1\,m^3$ is

$$S_{trans,N} = k\ \ln\frac{10^{30N}}{N!} \approx k\ \ln\frac{10^{30N}}{e^{-N}N^N} = k\ \ln\frac{e^N 10^{30N}}{N^N}$$
$$= kN\ \ln\frac{e\cdot 10^{30}}{N} = kN\left(\ln e\cdot 10^{30} - \ln N\right) \approx kN(70 - \ln N)$$

Since an increase in the number of molecules in the same space may increase the number of probable incidences of energy redistribution in the space, concentrated molecules may have more chances to redistribute their kinetic energy and have larger microstates if the temperature is kept the same. When two ideal gases are mixed under constant temperature in an isolated system, the total internal energy does not change during mixing while the total entropy increases, since both gases get more space to travel and have more chances to redistribute their kinetic energy.

Example 2.2 Calculate the root-mean-square speed, v_{rms} of oxygen at $T=293\,K$. The molecular mass of O_2 is $M=0.032\,kg/mol$. N_A is denoted by Avogadro's number. Oxygen is dealt with like an ideal gas at low concentration, and internal energy is proportional to temperature. Then the average kinetic energy of gas molecules is expressed as $3/2\ nRT$ [1,3].

$$U_{\text{trans}} \approx \langle E_{\text{trans}} \rangle = \frac{3}{2} nRT$$

$$\frac{1}{2} m N_{\text{A}} (v_{\text{rms}})^2 = \frac{3}{2} RT$$

$$M (v_{\text{rms}})^2 = 3RT$$

$$(v_{\text{rms}})^2 = \frac{3RT}{M} = \frac{3\left(8.314\,\text{J}\,\text{K}^{-1}\,\text{mol}^{-1}\right) \cdot 293\,\text{K}}{0.032\,\text{kg}\,\text{mol}^{-1}} = 22.8 \times 10^4\,\text{m}^2/\text{s}^2$$

$$v_{\text{rms}} = 478\,\text{m/s}$$

At the same temperature, a gaseous molecule with a small molecular mass like hydrogen has rms speed higher than that of heavier molecules. However, the average speed of the same molecular weight of molecules only depends on the temperature.

2.6 Diffusion of interacting molecules

Diffusion begins when solids dissolve in liquid, different gases or liquids mix, solutes such as sodium ions spread in the fluid of the axon when they flash in, or gases are dissolved in the blood. An interacting mass of gas molecules has the potential to diffuse spontaneously if they can obtain a sparser space. Liquid molecules have the microstates and entropy that are mainly associated with interaction like gas molecules but are different from gas molecules in that liquid molecules fill the space and bump continuously exchanging their energy and positions. Diffusion begins only when the concentrated area is made by an injection of a solute into a solution in a container or by putting different kinds of liquids together in the same container. In a highly concentrated spot of a certain kind of molecule, the pressure produced by bumping between them pushes one molecule away, while another kind of adjacent molecule gets into the concentrated area of certain molecules, thereby causing the diffusion by mixing. The pressure of molecules drives mixing until concentration becomes the same anywhere in the container under constant temperature and pressure (see Section 2.6.1). The speed of diffusion is proportional to the concentration gradient of a certain kind of molecules. A higher concentration gradient suggests a higher probability of a collision between the same kind of molecules causing a replacement of one of the positions with another kind of adjacent molecule, thereby causing the diffusion by fast mixing. The speed of mixing is fast at first and gradually goes down. The speed of ion diffusion is proportional to how much the the same kind of ion is concentrated in an area of the tissue, and it is inversely proportional to the resistivity of the tissue. Fick's law of diffusion is used to interpret the mixing and dilution of solutes dynamically and is applied to solve currents of ions (see Sections 5.2 and 5.7).

If the ion resistivity of the tissue is extremely high, the concentrated ionic molecules may bump and travel zigzag through it due to pressure fluctuation, which may make them oscillate. Diffusion of ions through a leak channel from the sarcoplasmic reticulum (SR) and injected ions in a model axon shows the oscillations which will not be seen in sparse gaseous molecules in a large space. The oscillation of concentrated ions may cause the oscillation of electrical potential in living nerve and muscle fibers in an exciting state. By solving the cable equation, we found that the membrane potential in the excited model axon shows the damped oscillation and withers over time and a distance that measures how far away from the origin of the current injection (see Section 6.9).

Question 2.3 (1) Why do gas molecules have huge energy microstates? (2) Why can we not predict that the entropy of a large polyatomic molecule is larger than that of a small one? (3) Why increasing temperature of liquid or gas molecules causes their faster diffusion? (4) Is the energy distribution to any independent oscillators in a liquid molecule and the microstate distribution to interacting liquid molecules near equilibrium the same?

Answer

(1) Gaseous molecules behave quite differently from solids and liquid molecules in that each molecule is distributed in a large space and possesses different amounts of kinetic energy in equilibrium. The number of microstates associated with the interaction can be calculated statistically using the lattice model. The number of probable energy distribution attributed to redistributing the kinetic energy between the molecules when they collide in a space becomes huge. It depends on the size of a space, the number of molecules, and the temperature.

(2) Generally, a large polyatomic molecule has more energy acceptors like oscillators and can hold more energy than a small one at the same temperature. That is why the entropy of a large polyatomic molecule is usually larger than that of a small one at the same temperature. However, a light

smaller molecule with a small intermolecular attraction tends to be liquid or gas at room temperature and obtains much larger entropy than a solid does due to additional huge microstates attributed to redistribution of their kinetic energy between molecules in the space they occupy.

(3) When the temperature is high, molecules have a larger kinetic energy which generates higher pressure when they collide.

(4) No, they are different. In an isolated interacting system, all interacting solids may have about the same average energy near equilibrium where microstates of the solids appear to be the maximum. The energy distribution to any independent oscillators in a solid in equilibrium follows the Boltzmann distribution while the microstate distribution to interacting solids near equilibrium follows the Gaussian distribution (see Section 2.3). The same thing is true for liquid molecules filling the space, interacting with each other, and exchanging energy and positions. As a result of the interaction, each of them can receive about the same amounts of average kinetic energy under isothermal equilibrium conditions where the microstates become the maximum. The distribution of energy to any independent oscillators in a liquid molecule in equilibrium follows the Boltzmann distribution while the distribution of microstates to interacting liquid molecules near equilibrium follows the Gaussian distribution.

2.6.1 Mixing of two pure liquids

If a liquid is kept in a container where molecules exchange kinetic energy randomly every time they collide, molecules exchange positions but are not able to diffuse. Molecules in a container may have about the same average kinetic energy due to continuous bumping with others under isothermal conditions. Entropy is kept the same under constant temperature. The diffusion called current will occur if we give more space to spread. A liquid is different from gas in that the space is fully occupied by molecules, and the rate of diffusion is slower than that of gas. Bumping between molecules produces pressure which drives them to diffuse. If two different kinds of liquid molecules are put together in a container, the total volume of a space increases, and the pressure produced by bumping between the same kind of molecule (A) drives them to push one away, while adjacent another kind of B molecule gets into the concentrated area of A molecules and replace A molecule with B molecule, thereby causing the diffusion by mixing. The pressure of molecules drives mixing until the concentration becomes the same anywhere in the container. We can see this phenomenon during gas exchange between the blood in artery and tissues (see Chapter 9). Mixing makes entropy larger due to an increase in the number of probable incidences of redistribution of kinetic energy related to exchanging positions between different kinds of molecules under isothermal conditions. The diffusion by mixing will stop when its concentration becomes the same anywhere in the container, where entropy becomes maximized.

The diffusion by mixing of two different kinds of liquid molecules separated in the container, starts with the removal of the separation film. Right after removal of the separation film, two molecules in the right side bump molecules in the left side and try to get in the left side while two molecules in the left try to get in the right, thereby causing the diffusion by mixing as shown in (1) of Fig. 2.3. The pressure by bumping pushes the four molecules in the right side to the left side while three molecules in the left side get in the right side as shown in (3) of Fig. 2.3. Exchanging of positions between the same kind of molecules occurs more often as concentration gradient goes down over time, thereby reducing the speed of diffusion as shown in (4) of Fig. 2.3. The speed of diffusion is proportional to the concentration

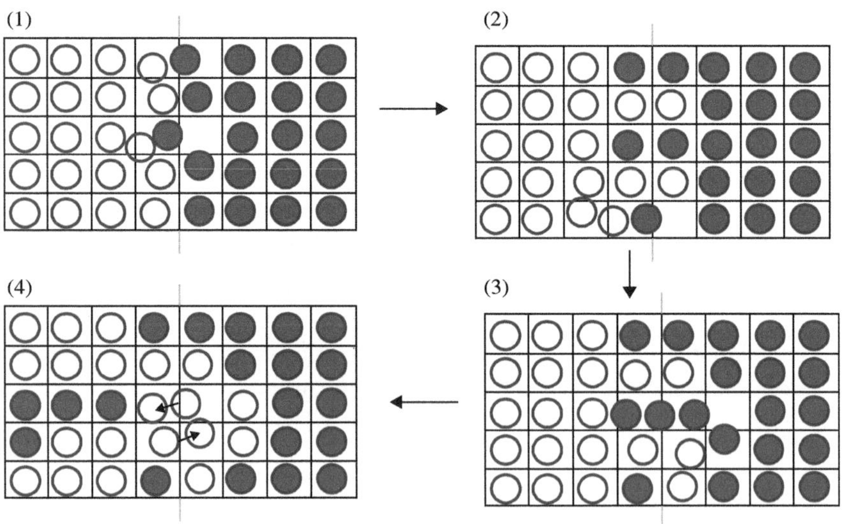

Fig. 2.3 (1) Right after removal of the separation film, two molecules in the right side bump molecules in the left side and try to get in the left side while two molecules in the left try to get in the right, thereby causing the diffusion by mixing. (3) The pressure by bumping drives the four molecules in the right side to the left side while three molecules in the left side get in the right side. (4) Exchanging of positions between the same kind of molecules occurs more often as concentration gradient goes down over time, thereby reducing the speed of diffusion.

gradient of a certain kind of molecules since a higher concentration gradient will cause a higher probability of a collision made between the same kind of molecules that replaces one of the positions with a different kind of adjacent molecule. The speed of mixing is fast at first and gradually goes down.

Let us calculate the change in entropy statistically when two pure liquids, A and B are mixed under constant pressure and temperature. If heat transfer at constant pressure (q_p) is zero, ΔH_{mix} of mixing is zero, indicating no bonds are formed or no intermolecular forces are exerted between two pure molecules.

$$q_p = 0, \Delta H = q_p = 0$$

The lattice model of a liquid is different from that of gas in that the lattice space is fully occupied by the molecules. The bumping pure molecules have the maximum microstates in equilibrium (see Section 2.3), and the maximum states are kept unchanged until mixing occurs where molecules have about the same kinetic energy and exchange positions by pushing. An increase in entropy by diffusion begins when the mixing starts. The number of maximum states of pure liquids is set to zero as the initial condition (Fig. 2.4). We will count only the change in entropy due to the exchanging positions by two different kinds of molecules.

N_{A} is the number of A molecules. N_{B} is the number of B molecules. The initial condition of entropy before mixing is set to zero as follows.

$$S_{\mathrm{A}} = k \ln \Omega_{\mathrm{A}} = k \ln \frac{N_{\mathrm{A}}!}{N_{\mathrm{A}}!} = 0$$

$$S_{\mathrm{B}} = k \ln \Omega_{\mathrm{B}} = k \ln \frac{N_{\mathrm{B}}!}{N_{\mathrm{B}}!} = 0$$

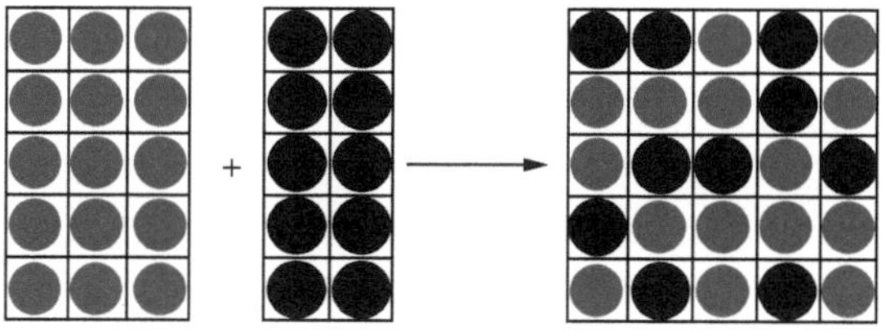

Fig. 2.4 Two pure ideal liquids are mixed under constant pressure and temperature. Molecules diffuse and exchange positions driven by the kinetic energy until the two liquids are thoroughly mixed. Two different kinds of molecules are assumed to have about the same molecular size. The number of maximum states of pure liquids is set to zero as the initial condition. We will count only the change in entropy due to the exchanging positions by two different kinds of molecules.

The size of two different kinds of molecules is assumed to be about the same. N is the total number of A and B molecules. The signs of X_A and X_B denote a fraction of A and B, respectively. An increase in the number of probable energy redistributions over molecules due to exchanging positions by two different kinds of molecules begins with mixing, which relates to an increase in the total number of probable locations they can take. The number of probable locations the same species of molecule can take in a space is considered as overcounting since exchanging positions among the same species of molecules cause no mixing and occur before and after the mixing. To correct the overcounting, we must divide by the number of ways of arranging molecules A and B, that is, $N_A!\,N_B!$. The total kinetic energy remains the same, but the number of probable energy states increases during mixing. It is calculated statistically using the lattice model.

$$\Omega_{mix} = \frac{N!}{N_A!N_B!}$$

The change in entropy after the mixing is

$$\Delta S_{mix} = S_{mix} - (S_A + S_B) = k\ \ln \Omega_{mix} = k\ \ln \frac{N!}{N_A!N_B!}$$

$$\ln(N!) \approx N \cdot \ln N - N \text{ when } N \text{ is very large.}$$

$$\begin{aligned}\Delta S_{mix} &= Nk \ln N - Nk - (N_A k \ln N_A - N_A k + N_B k \ln N_B - N_B k) \\ &= (N_A + N_B)k \ln N - N_A k \ln N_A - N_B k \ln N_B \\ &= N_A k\ \ln \frac{N}{N_A} + N_B k\ \ln \frac{N}{N_B}\end{aligned}$$

$$\text{Since } N_A = NX_A, N_B = NX_B, \text{and } 1 > X_A > 0, 1 > X_B > 0,$$

$$\Delta S_{mix} = -Nk(X_A \ln X_A + X_B \ln X_B) > 0 \tag{2.11}$$

Entropy is always increasing during mixing. Real liquids are considered to have additional microstates which are associated with molecular vibration and rotation, but these additional states occur in both the pure and the mixed liquid [4]. ΔS_{mix} is as a result of an increase in microstates by diffusion. This statistical justification is consistent with the result from the classical thermodynamics.

If ΔH_{mix} is almost zero, $(dS)_{H,\ p} > 0$ is the condition for spontaneity under constant H and the external pressure (see Section 2.6.3).

2.6.2 Dilution of solution

Changes in entropy and Gibbs free energy by dilution of a solution can be evaluated statistically if the solute and solvent are not attractive and aggregate among them, and their size is assumed to be about the same.

$S = k\,ln\Omega$, and $\Delta S = k\ \ln\frac{\Omega_f}{\Omega_i}$ where Ω_f is the number of final energy states per solute molecule after dilution, and Ω_i is the number of initial states per solute molecule before dilution.

The number of states per a solute has been known to be inversely proportional to its concentration (see Question 2.4).

For dilution of 1 mol of molecules,

$$\Delta S_{dil} = kN_A\ \ln\frac{\Omega_f}{\Omega_i}$$

where N_A is Avogadro's number.

$$\Delta S_{dil} = R\ \ln\frac{[\text{initial concentration}]}{[\text{final concentration}]} > 0 \qquad (2.12)$$

If $\Delta H_{dil} \approx 0$, $(\Delta S)_{H,\,p} > 0$ is the condition for spontaneity under constant pressure.

Under constant temperature T, let us calculate ΔG_{dil} for 1 mol. For dilution of 1 mol of solute molecules,

$$\Delta G_{dil} = \Delta H_{dil} - T\Delta S_{dil} = -T\Delta S_{dil} = -RT\ \ln\frac{[\text{initial concentration}]}{[\text{final concentration}]} \qquad (2.13)$$

The dilution process proceeds spontaneously under constant temperature and pressure because ΔG_{dil} is negative if heat transfer at constant pressure (q_p) is zero, i.e., ΔH_{dil} is about zero.

Question 2.4 Why is the number of states per a small solute molecule inversely proportional to its concentration?

Answer All solute and solvent molecules are assumed to be identical. Let the molecular volume be v and the total volume be V.

An increase in the number of microstates due to dilution is calculated statistically using the lattice model. The number of probable locations molecules can take in the space provides the number of locations to exchange positions and relates to the number of probable redistributions of the kinetic energy which causes consecutive new microstates. The number of probable locations the first of N molecules can take is $\Omega_{trans,1} = V/v$, that for the second is $\Omega_{trans,2} = (V/v) - 1$, that for the third is $\Omega_{trans,3} = (V/v) - 2$, and so on (Fig. 2.5). The number of probable locations the same species

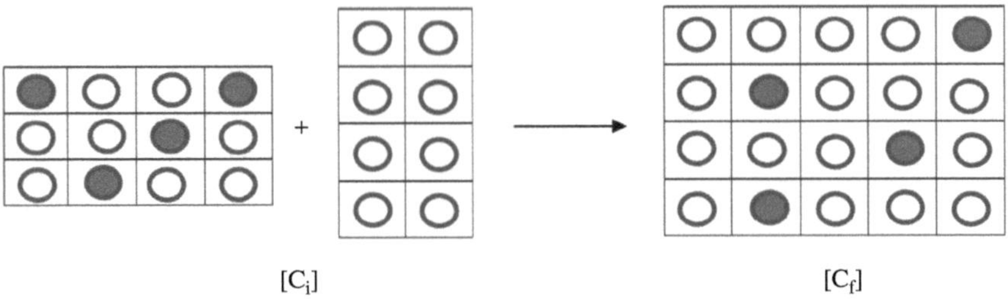

Fig. 2.5 Dilution of a solution under constant H, and under constant pressure where the internal pressure is in equilibrium to the external pressure. Volume changes from V_i to V_f and concentration changes from $[C_i]$ to $[C_f]$.

of molecules can take in a space is considered as overcounting since exchanging positions among the same species of molecules cause no diffusion. The total number of microstates for N solute molecules in the space, V is divided by $N!$

$$\prod_{i=1}^{N}\Omega_{\text{trans}} = (V/v)\{(V/v)-1\}\{(V/v)-2\}\cdot\cdots/N! \approx (V/v)^N/N!$$

Entropy of the solute increases as volume changes from V_i to V_f and concentration changes from $[C_i]$ to $[C_f]$.

$$\Delta S_{\text{dil},N} = k\ln\Omega_{N,\text{f}} - k\ln\Omega_{N,\text{i}} = k\ \ln\frac{\Omega_{N,\text{f}}}{\Omega_{N,\text{i}}} \approx k\ \ln\frac{(V_\text{f}/v)^N/N!}{(V_\text{i}/v)^N/N!}$$

$$\frac{\Delta S_{\text{dil},N}}{N} = \frac{k}{N}\ \ln\frac{\Omega_{N,\text{f}}}{\Omega_{N,\text{i}}} = k\ \ln\left(\frac{\Omega_\text{f}}{\Omega_\text{i}}\right) \approx k\ \ln\frac{V_\text{f}}{V_\text{i}} = k\ \ln\frac{\frac{N}{V_\text{i}}}{\frac{N}{V_\text{f}}} = k\ \ln\frac{[\text{C}_\text{i}]}{[\text{C}_\text{f}]} > 0 \tag{2.14}$$

The number of states per a small solute is inversely proportional to its concentration.

Entropy increases as the concentration of a solute molecule decreases. Entropy is always increasing during dilution, where there is a concentration gradient. Entropy is maximized when the concentration gradient becomes zero. Changes in entropy occur when the total volume increases and concentration decreases.

If heat transfer at constant pressure (q_p) is zero, i.e., $\Delta H \approx 0$, $(dS)_{H,\,p} > 0$ is the condition for spontaneity under constant external pressure (see Section 2.6.3).

Dilution goes spontaneously under constant external pressure until the concentration gradient disappears if no bonds are formed or no intermolecular forces are exerted during the process.

2.6.3 Spontaneity

In a reversible transformation, the system departs infinitesimally from equilibrium and remains at equilibrium throughout the change. The difference between the internal pressure, p_{in} and the external pressure, p_{ext} is infinitesimal.

Under reversible condition, $TdS = dq_{rev}$ by the definition [5].

In the reversible process, the change in internal energy is,

$$dU = TdS - p_{in}dV \tag{2.15}$$

However, real changes in our life are irreversible or spontaneous changes which depart from equilibrium throughout the changes. Spontaneous processes are always irreversible. The irreversible work done by a system to the surrounding is negative and expressed as $-p_{ext}dV$. The internal pressure, p_{in} is different from the external pressure, p_{ext}. Heat (q_{irrev}) should be supplied for the system to do continuous work.

$$\text{Irreversible Process:} \qquad dU = dq_{irrev} - p_{ext}dV$$

The difference in the internal energy only depends on the initial state (T_1, V_1) and the final state (T_2, V_2) of the system, but it does not depend on whether the process is reversible or irreversible.

$$dU = dq_{irrev} - p_{ext}dV = TdS - p_{in}dV$$

$$dV > 0, \quad p_{in} > p_{ext} > 0, \quad 0 > -p_{ext}dV > -p_{in}dV$$

$$\text{Therefore, } TdS > dq_{irrev}.$$

The condition placed on an irreversible change is

$$dU = dq_{irrev} - p_{ext}dV < TdS - p_{ext}dV$$

Another form of the condition on an irreversible change can be written as

$$dU + p_{ext}dV - TdS < 0 \tag{2.16}$$

(1) If $q = w = 0$ in an isolated system like a thermos bottle, $dU = 0$ and $w = p_{ext}dV = 0$.

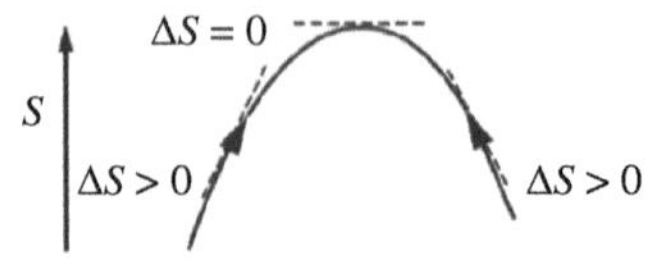

Fig. 2.6 In an isolated system, the process goes continuously toward the maximum entropy. Equilibrium is achieved when entropy is maximized.

$(dS)_{U,\ V} > 0$ is the condition for an irreversible process for an isolated system according to Eq. (2.16).

Since a spontaneous process is always irreversible, $(dS)_{U,\ V} > 0$ is the condition for spontaneity for an isolated system. The process goes continuously until equilibrium is achieved when entropy is maximized (Fig. 2.6).

(2) If H is constant and the external pressure is constant, $dH = dU + p_{\text{ext}} dV = 0$.
$(dS)_{H,\ p} > 0$ is the condition for spontaneity under constant H and constant external pressure.

(3) *Gibbs free energy* is defined as $G \equiv H - TS$.

$$G = H - TS = U + pV - TS$$

If the system changes under constant T and external pressure, p_{ext},

$$(dG)_{T,p} = dU + p_{\text{ext}} dV - TdS$$

$(dG)_{T,\ p} < 0$ is the condition for spontaneity under constant external pressure and temperature. At constant external pressure and temperature, equilibrium is achieved when the Gibbs free energy is minimized (Fig. 2.7).

2.6.4 Aggregation of lipid molecules in water

When phospholipids are dissolved in water, the hydration shell of separated phospholipid molecules requires more fixed water molecules than the hydration shell of the aggregated phospholipid molecules. The aggregation of phospholipids in water decreases the fixed water molecules around them

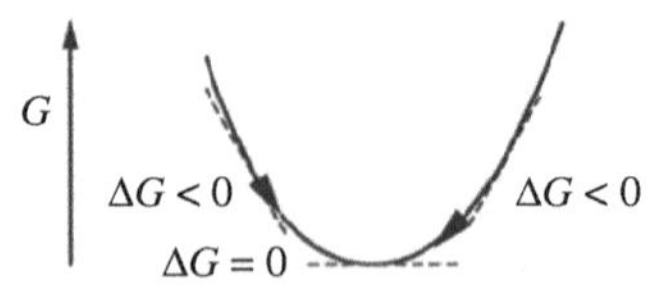

Fig. 2.7 At constant external pressure and constant temperature, the process goes continuously toward the minimum Gibbs free energy. During the process, the external pressure is thought to be in equilibrium to the internal pressure. Equilibrium is achieved when Gibbs free energy is minimized.

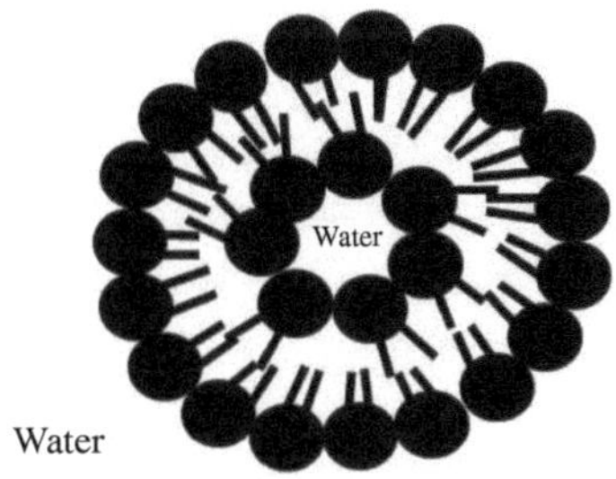

Fig. 2.8 Liposome is a lipid vesicle made of a concentric bilayer of lipids with polar heads attached to hydrophobic tails. The aggregation in water occurs spontaneously due to an increase in entropy and the exothermic process attributable to the van der Waals force.

and increases the free water molecule in the solution. Since entropy of the fixed water molecule is smaller than entropy of the free water molecule, the aggregation increases entropy. When the van der Waals force attracts hydrophobic tails, energy is released, which means aggregation is an exothermic process. Aggregation of the nonpolar part of molecules in an aqueous solution with an exclusion of water molecules is called the hydrophobic effect (Fig. 2.8).

The aggregation of phospholipids in water occurs spontaneously due to an increase in entropy and the exothermic process attributable to the van der Waals force.

$$(dG)_{T,p} = dH - TdS < 0$$

$(dG)_{T,\ p} < 0$ is the condition for spontaneity under the constant surrounding pressure and temperature (see Section 2.6.3).

2.7 Conclusions

Entropy, as defined by Boltzmann, is carefully derived statistically as a function of ln Ω. The number of microstates (Ω) is the number of probable energy distributions over oscillators in the solid. In the isolated system at equilibrium, the system has the maximum entropy, and then the Boltzmann distribution appears. The microstates and entropy associated with molecular interaction increase every time molecules collide, exchange kinetic energy in the space, and finally stop at the maximum when equilibrium is attained in an isolated system. The number of locations molecules probably take in a space provides the number of locations where they exchange positions and relates to the number of probable redistributions of the kinetic energy, which determines the number of microstates on molecular interaction. Huge microstates and entropy of gas are mainly as a result of the molecular interaction associated with an exchange of their kinetic energy occurring every time they collide in an isothermal space. An interacting mass of gas molecules has the potential to diffuse spontaneously if they can obtain a sparser space. Liquid molecules have the microstates and entropy that are mainly associated with interaction like gas molecules but are different from gas molecules in that liquid molecules fill the space and bump continuously exchanging their energy and positions. The pressure produced by bumping between the same kind of molecule (A) drives them to push one away, while adjacent another kind of B molecule gets into the concentrated area of A molecules and replace A molecule with B molecule, thereby causing the diffusion by mixing. A higher concentration gradient suggests a higher probability of a collision between the same kind of molecules causing a replacement of one of the positions with another kind of adjacent molecule, thereby causing the diffusion by fast mixing. The speed of mixing is fast at first and gradually goes down. The diffusion by mixing will stop when its concentration becomes the same anywhere in the container, where entropy becomes maximized. We can see this phenomenon during gas exchange between blood and tissues.

References

[1] T. Watanabe, Biophysical Basis of Physiology and Calcium Signaling Mechanism in Cardiac and Smooth Muscle, Academic Press, San Diego, CA, 2018 (Chapter 2).
[2] V.S. Daniel, An Introduction to Thermal Physics, Addison Wesley Longman, San Francisco, CA, 2000 (Chapter 2).
[3] W.C. Gilbert, Physical Chemistry, third ed., Addison-Wesley, Reading, MA, 1983 (Chapter 4).
[4] K.A. Nelson, M. Bawendi, Thermodynamics & Kinetics, MIT OpenCourseware, 2008 (Lecture # 24-28).
[5] W.C. Gilbert, Physical Chemistry, third ed., Addison-Wesley, Reading, MA, 1983 (Chapter 10).

CHAPTER 3

Function of membrane and protein synthesis in the cell

3.1 Introduction

The cell membranes are made of phospholipids, cholesterol, and proteins. The phospholipid bilayer forms the basic structure of the membrane. This self-orienting property of phospholipids leads the biological membrane into a closed spherical structure and reseals it by itself when torn out. Only gases such as oxygen, carbon dioxide, and nitrogen, and small nonpolar lipid-soluble molecules can diffuse freely through the phospholipid bilayer membrane along their concentration gradients. However, ions and polar molecules cannot diffuse through it, and their flux through the membrane requires channels or transporters that are made of columnar proteins firmly inserted into the phospholipid bilayer. Each kind of channel has a pore allowing the passage of only selected solute. Some channels are always open such as aquaporins (water channels) and potassium leak channels. But some channels have two states, open and closed, which can be controlled by bonding the ligand to the receptor on the channel protein, or by means of the voltage or mechanical sensor. On the other hand, a transporter forms an intermediate complex with the solute, and a subsequent conformational change in the transporter causes translocation of the solutes to the other side of the membrane. Passive transport processes require no energy supply. They are designed for the diffusion of specified polar molecules such as glucose and amino acids that are too large to go through membrane channels by themselves. Active transport processes are used for ions to diffuse across cell membranes against the electrochemical gradient by means of energy supplied by ATP hydrolysis. Action of Na^+-K^+ ATPase is an important example of active transport. The ion flux through channels or transporters causes ionic current and can change the electric potential inside the cell.

Mitochondrial DNA is bacteria-like DNA. Aerobic bacteria engulfed are thought to evolve into mitochondria and provide the eukaryotic cells with the ability to synthesize ATP. F_0F_1 ATP-synthase is the rotating

Modeling Electrochemical Dynamics and Signaling Mechanisms in Excitable Cells with Pathological Case Studies
https://doi.org/10.1016/B978-0-323-98803-2.00011-0

transmembrane protein driven by proton flux between the intermembrane space and matrix in a mitochondrion.

The functions of living cells are regulated by charged or polar signaling small molecules called ligands binding to the receptors. Some ligands can change the conformation of the ion channels in the cell membranes when they specifically bind to the receptors and open the channels for specific ions. The selective permeability of membrane creates osmosis, and interestingly with the help of continuous actions of Na^{+}-K^{+} pump, it causes the membrane potential.

3.2 Molecular shape of proteins

Proteins and peptides are amino acid polymers. An amino group from one amino acid forms an amide bond with the carboxyl of a second amino acid. There are two types of secondary structure. The alpha helix is a right-handed coil. The R groups of residues extend outward from the peptide backbone of the helix. The coiling results from the hydrogen bonding within a protein molecule between the electronegative carbonyl oxygen atom of one amino acid residue and the electropositive hydrogen atom of the N-H group of another. The alpha helical secondary structure is common in the fibrous structural proteins called keratins, which make hair and transmembrane columnar protein. A beta-pleated sheet is formed from more t-han two peptide chains that are extended and lying next to one another. A beta-pleated sheet is seen between different regions of the same peptide where the peptide chain is bent back. The sheet is stabilized by hydrogen bonds between the hydrogen of the N-H groups on one chain and the carbonyl oxygen on the other neighboring chain. Many proteins contain regions of both alpha helix and beta-pleated sheets in the same polypeptide chain.

Some amino acids have hydrophobic side chains and others have hydrophilic side chains. In the aqueous environment, the hydrophobic core regions of the protein move inside while hydrophilic polar side chains face outside to form a hydrogen bond with polar molecules of water and charged side chains facing outside attract some ions and some water molecules. Hence, an unfolded protein becomes globular in water. Changes in temperature or pH will break its tertiary structure and make denatured proteins whose primary structure remains the same, but the conformational changes of tertiary structure occur. Unfolding a globular shape of protein results in its elongated looped chain.

Transmembrane proteins are firmly inserted into the phospholipid bilayer. They have both hydrophilic and hydrophobic regions. Spanning the entire width of the membrane, a series of negatively charged hydrophilic side chains and carbonyl groups of the residues protrude from both sides of the cell membrane and the inside of ion channels, interacting with water and positive ions, while the hydrophobic regions interact with phospholipid tails and cholesterols inside the membrane.

Once the amino acid sequence of the protein was determined, computer programs can estimate the shape of the protein.

Amino acids which have nonpolar chains are glycine (Gly), alanine (Ala), valine (Val), leucine (Leu), isoleucine (Ile), methionine (Me), phenylalanine (Phe), tryptophan (Trp), and proline (Pro).

Amino acids which have a polar side chain are serine (Ser), threonine (Thr), cysteine (Cys), tyrosine (Tyr), asparagine (Asn), glutamine (Gln).

Amino acids which have negatively charged side chains are aspartic acid (Asp), glutamic acid (Glu).

Amino acids which have positively charged side chains are lysine (Lys), arginine (Arg), histidine (His).

3.3 Phospholipid bilayer cell membrane

The phospholipid bilayer forms the basic structure of the membrane. Each phospholipid molecule has a polar head that is charged and two nonpolar hydrocarbon tails from fatty acids. A polar head is hydrophilic. It is so-called since it attracts polar water molecules. Nonpolar tails are hydrophobic, which avoid water and align in the center of the two-layer membrane. The polar heads are exposed to water both inside and outside of the cell. The nonpolar tails lie between the polar heads making a sandwich-like structure. This self-orienting property of phospholipids in aqueous solution leads the biological membrane into closed spherical structures and reseals it by itself when torn out (see Section 3.3.1). The transmembrane proteins are firmly inserted into the phospholipid bilayer. Spanning the entire width of the membrane, the negatively charged hydrophilic parts of the protein protrude from both sides of the cell membrane where polar branching glycans are located, while the hydrophobic parts of the residues interact with lipid tails and cholesterols inside the membrane. Therefore, the permeability of membrane to small nonpolar molecules can be modulated by hydrophobic material properties inside the membrane [1].

3.3.1 Liposome

Liposomes were first described in 1961 at the Babraham Institute, in Cambridge. They were discovered when Bangham and R.W. Horne were testing the new electron microscope. The microscope pictures served as the first evidence for the cell membrane being a bilayer phospholipid structure. When phospholipids are dissolved in water, van der Waals force attracts the hydrophobic tails of the lipids, and aggregation occurs spontaneously. The aggregation of phospholipids in water decreases the fixed water molecules around them. Aggregation of the nonpolar part of molecules in an aqueous solution with exclusion of water molecules is called the hydrophobic effect.

Liposome is a lipid vesicle made of a concentric bilayer of phospholipids with polar head groups attached to hydrophobic tails and considered the simplest form of a biological cell. It can be artificially made and used to deliver microscopic substances such as negatively charged nucleic acids into body cells that have a negatively charged membrane (Fig. 3.1). When some of the cationic liposomes containing nucleic acids are injected, they are attracted by the negatively charged cell membrane and can be either engulfed by the cell membrane or fuse with the cell membrane. Some of the nucleic acids will enter the host cell. If the nucleic acid is DNA, it needs to get into the nucleus for its effect. This happens when the cell undergoes mitosis. The nuclear membrane breaks down and some DNA gets trapped in the nuclei of the newly formed daughter cells.

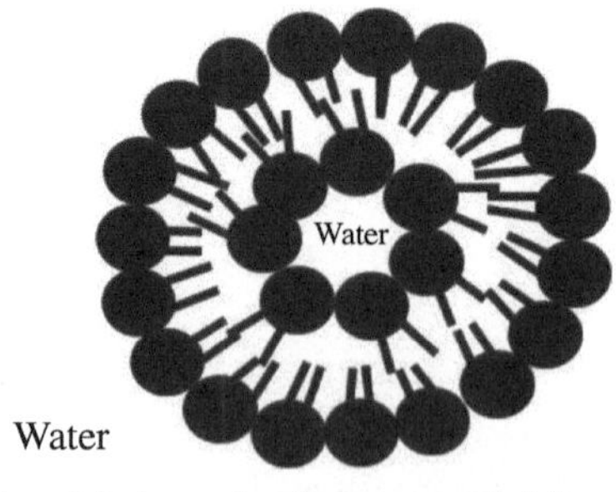

Fig. 3.1 Liposome, a lipid vesicle is made of a concentric bilayer of lipids with polar head groups attached to hydrophobic tails and considered the simplest form of the biological cell and can be used to deliver microscopic substances to body cells.

3.3.2 Molecular structure of the membrane

The majority of membrane phospholipids have one unsaturated nonpolar hydrocarbon chain which kinks their tails whereby increasing the space between them and the membrane fluidity as shown in Fig. 3.2.

Cholesterol fills the space created by the phospholipid tails. The hydroxyl group of cholesterol interacts with the polar head group of membrane phospholipids, while the tetracyclic carbon skeleton and the hydrocarbon chain are embedded in the membrane alongside the nonpolar tails of the phospholipids. The bulky structure of the tetracyclic ring of cholesterol contributes to add some rigidity to the cell membrane along with proteins. It changes membrane quality of fluidity and enables animal cells to protect membrane integrity while changing shape and moving about. In this structural role, cholesterol reduces the permeability of the cell membrane to polar solutes and ions. Transmembrane proteins are firmly inserted into the phospholipid bilayer. Many of the transmembrane proteins which abut the extracellular fluid are glycoproteins, and they strengthen the membrane. They give not only an extra mechanical strength to the membrane but also make the cell capable to communicate with the other cell.

3.3.3 Disorientation of proteins and disintegration of membrane lipids

In water, van der Waals forces attract the hydrophobic residues, and hydrophobic effects can stabilize the close interactions among them. The hydrophobic core regions of the protein move inside while polar side chains and carbonyl groups of amino acid residues face outside to form a hydrogen bond with polar molecules of water, and the charged side chains facing outside

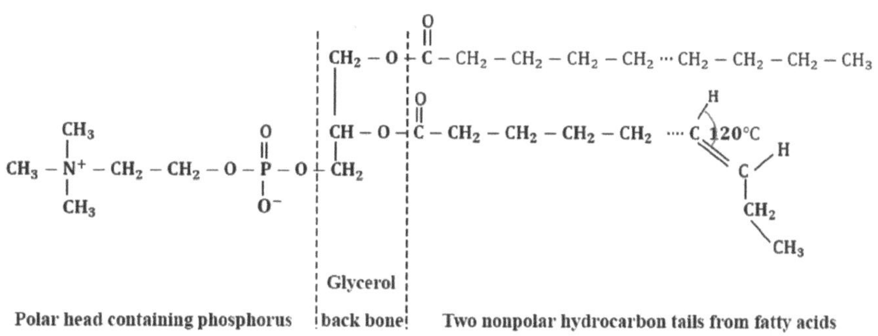

Fig. 3.2 Phosphatidylcholine is cationic phospholipid molecules. The phospholipid has one unsaturated nonpolar hydrocarbon chain that kinks their tails thereby increasing the space between them.

attract some ions and some water molecules. Hence, a folded protein becomes globular and solved in water. Chaotropic agents will break its tertiary structure and make denatured proteins whose primary structure remains the same, but the conformational changes of tertiary structure occur. The chaotropic solutes such as salts [2], isopropyl alcohol, methanol, and guanidium thiocyanate increase entropy of the system by interfering with intermolecular interactions caused by electrostatic forces such as electrostatic interaction between a hydrogen atom in a polar bond and a hydrogen acceptor (hydrogen bond) and between an instantaneous localization of electron on atoms of one side of a nonpolar large molecule and the neighbor molecule (van der Waals force).

The chaotropic agents can disrupt the solvation shell along with disruption of the hydrogen binding network between water molecules. Large ionic molecules such as DNA and RNA in aqueous solution tend to precipitate when salts and isopropyl alcohol or methanol are added. Isopropyl alcohol or methanol added to the aqueous solution disrupts the hydration shells. By adding salts, positively charged metal ions can neutralize the negatively charged phosphate backbone of RNA. The neutralized RNA begins to aggregate in the aqueous solution.

Besides that, the hydrophobic effect will be interfered by the chaotropic solutes. The chaotropic solutes will disintegrate membrane phospholipids by weakening van der Waals forces and hydrophobic effects. In addition, the chaotropic solutes enhance the solvation of the hydrophobic region inside of proteins by weakening van der Waals forces and hydrophobic effects along with disruption of the hydrogen binding network between water molecules. They reduce the amount of order in the structure of a protein and change the shape of globular protein and folded protein.

3.4 Bulk transport, endocytosis, and exocytosis

Endocytosis is a cellular process in which substances are taken into the cell. The cell membrane bends around, engulfs the material, and then the invagination buds off inside the cell to form a vesicle containing the ingested material. Endocytosis includes pinocytosis (drinking) and phagocytosis (eating). It is a form of active transport. Pinocytosis, which usually occurs from highly ruffled regions of the plasma membrane, is the invagination of the cell membrane to form a pocket. And then the pocket pinches off into the cell to form a vesicle (0.5–5 μm in diameter) filled with a large volume of extracellular fluid and molecules solved within it. The filling of the pocket occurs in a nonspecific manner. The vesicle then travels into the cytosol and fuses with other vesicles such as endosomes and lysosomes. Phagocytosis is the process by which cells bind and internalize particulate matter larger than around 0.75 μm in diameters, such as small-sized dust particles, cell debris, exsosomes, liposomes, nucleic acid-lipid complexes, microorganisms, and apoptotic cells.

Exocytosis is an active transport process in which a cell transports neurotransmitters and proteins out of the cell by secreting them through an energy-dependent process. Exocytosis is used by all cells because most chemical substances important to them are large polar molecules that cannot pass through the hydrophobic portion of the cell membrane by passive means. In exocytosis, membrane-bound secretory vesicle transiently fuses with the plasma membrane and then the area of fusion breaks releasing their water-soluble contents of the vesicle outside. Sudden influx of Ca^{2+} into the cell makes the fusion possible. In the neurotransmission, neurotransmitters are released from synaptic vesicles into the synaptic cleft via exocytosis. Exocytosis is also a mechanism by which cells can insert membrane proteins, lipids, and other components into the cell membrane. Vesicles containing these membrane components fully fuse with and become a part of the cell membrane.

3.5 Production, glycosylation, and transport of proteins

The ribosomes read messenger RNA (mRNA) and translate it into the protein. They are the sites of protein synthesis. A cytoplasmic protein such as hemoglobin is synthesized by the free ribosomes and need not be transported. However, all transmembrane proteins, secreted proteins such as insulin and glucagon, and membrane proteins stored in lysosomes need to be transported through the membrane. Since proteins cannot go through the membrane alone, the cell has a unique mechanism of the production of the membrane proteins and secreted proteins, and their transports in the cell. Messenger RNA and ribosomes meet in the cytosol for translation. When the signal peptide is synthesized, the signal peptide on the ribosome is attached to the endoplasmic reticulum (ER) to complete translation by putting the growing peptide chain in the ER. Vesicles containing the proteins formed in the ER are transported to the Golgi apparatus where the proteins are processed under glycosylation by adding, removing, or modifying hydrophilic glycans [3,4]. Some proteins cannot fold correctly unless they are glycosylated. In addition, the glycoproteins not only give extra mechanical strength to the cell membrane but also markers for cellular identification and make the cell capable to communicate with the other cell. From the Golgi, vesicles containig proteins shuttle to the lysosomes or to the cell exterior via exocytosis. In exocytosis, cytoplasmic sides of two membranes fuse and then the area of fusion breaks releasing contents of the vesicle outside. Every cell has transmembrane proteins with a specific pattern of branching glycan on the surface of a cell [5]. The specific carbohydrate on the cell surface makes cell-to-cell adhesion possible. Cells of the immune system can identify bacteria by adhering to and recognizing the specific pattern of branching glycan on the bacterial surface.

3.6 Columnar transmembrane proteins

Transmembrane proteins are firmly inserted into the phospholipid bilayer. Spanning the entire width of the membrane, they may protrude from both sides of the cell.

The columnar transmembrane proteins make up about half of the cell membrane by mass and are responsible for the specialized membrane functions. (1) Some transmembrane proteins are junctions connecting and joining two cells together. (2) Some transmembrane proteins have a specific pattern of branching glycan and may function as markers for cellular identification. (3) Some transmembrane proteins are aquaporins through which water can diffuse in and out according to the differences in water concentration between both sides of the membrane. (4) Some transmembrane proteins are pumps by which ions move across cell membranes against electrochemical potential using energy supplied by hydrolysis of ATP. (5) Some transmembrane proteins are passive transporter so that a specific small solute such as glucose and amino acids can diffuse through according to the concentration gradient. (6) Some transmembrane proteins are ligand-gated ion channels. Ligand-gated ion channels, also commonly called ionotropic receptors, are a group of transmembrane ion channel proteins. Neurotransmitters are signaling chemicals of the nervous system released from synaptic vesicles formed in the axon terminals of presynaptic neurons, bind specifically to the receptors on the ion channels in the postsynaptic membrane of the target cell such as a skeletal muscle fiber or a nerve cell, and the binding changes the conformation of the pore columnar protein and opens the channel for the specific ions such as Na^+, Cl^-, Ca^{2+}, and K^+ to be able to pass through it. Acetylcholine (ACh) at a neuromuscular junction that is a junction of a motor nerve axon terminal and a skeletal muscle fiber, ACh in an autonomic ganglion, i.e., a mass of nerve cell bodies at the junction of the first and second neuron of the autonomic nervous system, and others such as glutamate, serotonin (5HT), and gamma-aminobutyric acid (GABA) in the central nervous system are examples of neurotransmitters. (7) Some transmembrane proteins are voltage-gated Na^+, K^+, or Ca^{2+} channels. (8) Some transmembrane proteins are G protein-coupled receptors that detect ligands outside the cell and activate cellular responses by means of signal transduction pathways. They bind selectively to hormones such as epinephrine released from adrenal medulla, ACTH, and glucagon, or the neurotransmitters such as acetylcholine secreted from postganglionic parasympathetic nerves, norepinephrine released from postganglionic sympathetic nerves, and dopamine from dopaminergic nerves in the central nervous system, or other chemical messengers such as odorants. The binding of the ligands to the receptors relays signals down to the interior of cardiac and smooth muscle cells and operates cellular metabolic responses via changing activities of the enzymes which catalyze production of second messengers such as cyclic AMP in cardiac and smooth muscles, and IP_3 in smooth muscles [6], or opens potassium (GIRK, IKACh) channels located in the sinoatrial (SA) node in the heart via dissociated βγ-complex of G_i protein. (9) Some transmembrane proteins are enzymes such as F_0F_1 ATP synthase in a mitochondrion. When hormones such as insulin bind to the receptor, it activates the receptor's protein kinase protrusion facing the cytoplasm which activates a cascade of downstream signaling events in the cell. One of the important signals of insulin is to facilitate the metabolism of glucose by phosphorylation of glucose.

3.7 DNA and RNA

The nucleic acids are composed of carbon, oxygen, hydrogen, nitrogen, and phosphorus. The nucleic acids include two classes of molecules, deoxyribonucleic acid (DNA) and ribonucleic acid (RNA). They are the largest molecules in the body. The structural units of nucleic acids are called nucleotides. Each nucleotide consists of three components which are a nitrogen-containing base, a pentose sugar, and at least one negatively charged phosphate group. The nitrogenous bases are adenine (A), guanine (G), cytosine (C), thymine (T), and uracil (U). Adenine and guanine are two ring bases called purines, whereas cytosine, thymine, and uracil are single ring bases called pyrimidines. DNA and RNA are both polymers of nucleotides, but they differ in many respects. DNA is found in the cytoplasm of prokaryotic cells and in the nucleus and mitochondria of all eukaryotic organisms. Each chromosome contains an average of 80 million base pairs of DNA encoding several thousand genes. It replicates before the cell divides, which means the genetic information in the descendant cells is identical. The genetic code on DNA provides the basic instructions for building proteins in the body. A technique called DNA fingerprinting analyzes tiny samples of DNA taken from blood, or other body tissues, and shows the results as a genetic barcode that distinguishes each of us from others. DNA is a long, double-stranded polymer of nucleotides. The bases in DNA are A, G, C, and T, and its pentose sugar is 2′-deoxyribose (Figs. 3.3 and 3.4). How are these nucleotides combined? Nucleoside triphosphates (nucleotide containing three negatively charged phosphate groups) are polymerized by the DNA polymerase by copying a DNA template strand using base-paring interactions. The bond is called a

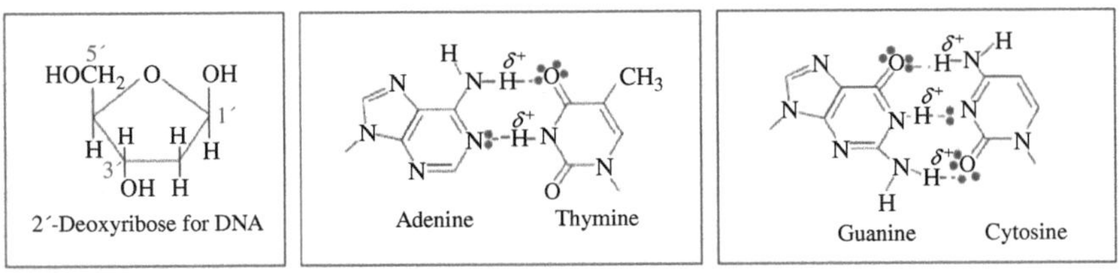

Fig. 3.3 Components of a nucleotide, a pentose sugar, and nitrogen-containing bases. The bases in DNA are adenine (A), guanine (G), cytosine (C), and thymine (T), and its pentose sugar is 2′-deoxyribose. The bases in RNA are adenine (A), guanine (G), cytosine (C), and uracil (U), and its pentose sugar is ribose. Hydrogen bonds between adenine and thymine, and between guanine and cytosine.

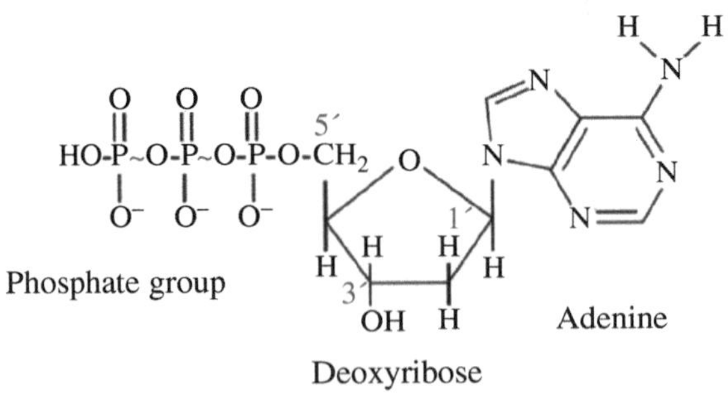

Fig. 3.4 This nucleotide is deoxyadenosine triphosphate (dATP) which has three negatively charged phosphate groups and a free OH group at 3′ of deoxyribose, and a very unstable anionic molecule. It is used for DNA synthesis as a substrate of DNA polymerase. Note ATP has OH group at 2′ position of the pentose ring.

phosphodiester bond between the 3′-hydroxyl group of deoxyribose and the 5′-phosphate. When the bond forms, the two phosphate groups leave. A new nucleotide is always added to the terminal of the growing strand at the free 3′-hydroxyl group of the deoxyribose.

Alternating sugar and phosphate components of each chain form the backbones of the ladder-shaped DNA (Fig. 3.5). Two chains of the backbones are held together horizontally by hydrogen bonds between bases. The purine A pairs with the pyrimidine T, and the purine G pairs with the pyrimidine C by base-paring interactions, respectively. The whole molecule is coiled into a spiral staircase-like structure called *double helix*. The two chains run in opposite directions.

Rosalind Franklin's work on the X-ray diffraction images of DNA and the alpha helical structure of a protein (Poling, Corey, 1951) led Francis Crick and James Watson proposed the model of the DNA double helix in 1953. According to Francis Crick, Franklin's data and research were the keys to determining the structure. James Watson affirmed this opinion later.

Contrary to DNA that is present in the nucleus and mitochondria of all eukaryotic organisms, RNA is located chiefly in the cytoplasm outside the

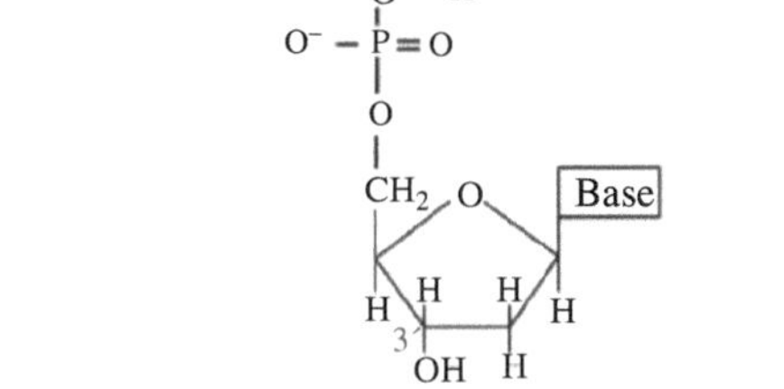

Fig. 3.5 Nucleotide containing three phosphate groups can be polymerized by DNA polymerase. The bond is called a phosphodiester bond between the 3′-hydroxyl group of deoxyribose and the 5′-phosphate. When the bond forms, the two phosphate groups leave. New nucleotide is always added to the growing strand at the free 3′-hydroxyl group of the deoxyribose by the polymerase. Nucleotides are held together by a strong phosphodiester bond using own high bond energy that each nucleoside triphosphate brings. Alternating sugar and phosphate components of each chain form the backbones of the ladder-shaped DNA.

nucleus. A gene is defined as a segment of a DNA molecule that carries instructions for one protein molecule. Messenger RNA (mRNA) carries out the code orders for the particular protein synthesis from the gene in the nucleus into the cytoplasm. Each code consists of three bases called a triplet which specifies the particular amino acid. The base sequence on mRNA is scanned and translated into a sequence of amino acids at ribosomal RNA (rRNA) in the small subunit of a ribosome. RNA molecules are a single strand of nucleotides which permit the formation of diverse structures that is straight or folded (Fig. 3.6). RNA bases include A, G, C, and uracil (U) that replaces the thymine (T) of DNA, and its sugar is ribose instead of deoxyribose. Unlike DNA, RNA molecules are unstable because of the presence of OH group at 2′ position of the pentose ring. Uracil has no methyl group that thymine has. MicroRNA molecules are abundant in many mammalian cells. They control genetic expression by shutting down or altering the expression. Recently, both mRNA and microRNA were discovered in exosomes that are present in many and perhaps all biological

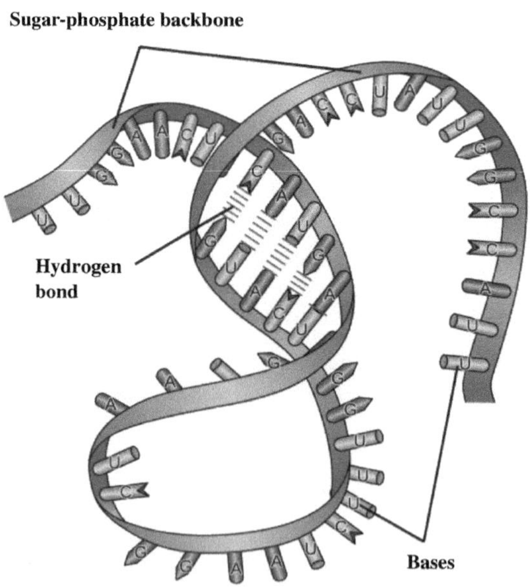

Fig. 3.6 Most RNAs are single-stranded, but some RNAs may contain regions that can form complementary base pairing where RNA strand loops back on itself resulting in some double-stranded regions. Ribosomal RNAs (rRNAs) and transfer RNAs (tRNAs) exhibit substantial secondary structure, as do some messenger RNAs (mRNAs).

fluids, including blood, urine, and the medium of cell cultures (see Section 3.12).

Question 3.1 Describe structural differences between DNA and RNA.

Answer (1) RNA bases include A, G, C, and uracil (U) that replaces the thymine (T) in DNA, and its sugar is ribose instead of deoxyribose in DNA. Unlike DNA, RNA molecules are unstable because of the presence of OH group at 2′ position of the pentose ring. (2) RNA molecules are a single strand of nucleotides that permit the formation of diverse structures that is unfolded or folded instead of double-stranded helix in DNA.

3.8 Protein synthesis in living cells

Proteins are made up of amino acids. A gene is defined as a segment of a DNA molecule that carries instructions for one peptide chain. There are four kinds of bases, A, G, T, and C, on DNA. Ribosomal RNA in the small subunit of a ribosome scans a combination of three consecutive nucleotides in mRNA and specifies the amino acid. AAA is the code for phenylalanine, AGA for serine, and TAC for methionine. These rules are called genetic code. Protein synthesis involves three steps. The first step is transcription in which the genetic information on DNA is passed to RNA. The three-base sequence on RNA is called a codon. Eukaryotes require the second step, which is RNA processing, that removes introns, and bases are added at both ends for the RNA to leave the nucleus via a nuclear pore. Introns are noncoding segments that range from 60 to 100,000 nucleotides. In prokaryotes, genes contain no introns and no nuclear envelope, no removal of introns and no addition of bases are needed. The third step is a translation in which the information carried by mRNA is decoded by the ribosome to assemble proteins.

3.8.1 RNA polymerase

RNA polymerase (DNA-directed RNA polymerase) is an enzyme that synthesizes RNA from a DNA template. The synthesis of RNA requires DNA as a template, ensuring that the base composition of the RNA faithfully reflects that of the DNA. In addition, as in DNA synthesis, all four nucleoside triphosphates must be present. RNA polymerase locally opens the double-stranded DNA so that one strand of the exposed nucleotides can be used as a template for the synthesis of RNA, a process called transcription. *In eukaryotic cells*, activator proteins (some transcription factor proteins) bind to pieces of DNA called enhancers. The binding causes the DNA to bend, bringing them near the promoter. Other transcription factor proteins join the activator protein, forming a protein complex that binds to the promoter. This protein complex makes it easier for RNA polymerase to attach to the promoter before the polymerase can initiate the DNA unwinding at that position and start transcribing a gene (Fig. 3.7). The polymerase not only initiates RNA transcription, but also guides the nucleotides into position, facilitates attachment and elongation, and has intrinsic proofreading and replacement capabilities and termination recognition capability.

The polymerase produces RNA that is either coding for protein (messenger RNA) or noncoding RNA. At least four functional types of RNA exist: (1) The ribosomal RNA (rRNA) in the small subunit of a ribosome that scans mRNA in the 5′ to 3′ direction: Translation initiation in *prokaryotes* usually requires interaction between 16S rRNA in the small subunit and an mRNA sequence upstream of the initiation codon, i.e., the ribosome binding site sequence called Shine-Dalgarno sequence (see Section 3.8.3). (2) Transfer RNA (tRNA) transfers specific amino acid to growing polypeptide chains in the large subunit of the ribosome during translation: When the ribosome reads a three-base sequence on mRNA, it calls for the amino acid carried by the adaptor molecule of transfer RNA. Each tRNA brings each specified amino acid dissolved in the cytosol into the large subunit (see Fig. 3.9). (3) Catalytic RNA (ribozyme) in the large subunit of a ribosome functions as an enzyme: The ribozyme in the large subunit catalyzes peptide bond formation of the amino acids carried by tRNA. (4) MicroRNA (miRNA) regulates gene expression by inactivating mRNA: Genes coding for miRNA are contained in the nucleus and are transcribed by RNA polymerase II, which produce a premature regulatory RNA. The regulatory RNA has a typical hairpin loop structure and will become the final miRNA with regulatory function after several steps at processing (see Section 3.8.4).

3.8.2 Transcription

A promoter is a special DNA sequence that specifies where RNA synthesis starts, and which DNA strand serves as the template strand. A promoter has the sequence which is recognized by RNA polymerase. In *prokaryotes*, the binding of the polymerase to the promoter is usually repressed as a repressor is binding to the operator for control of the expression of a gene. As soon as an inducer binds to the repressor, the repressor is detached from the

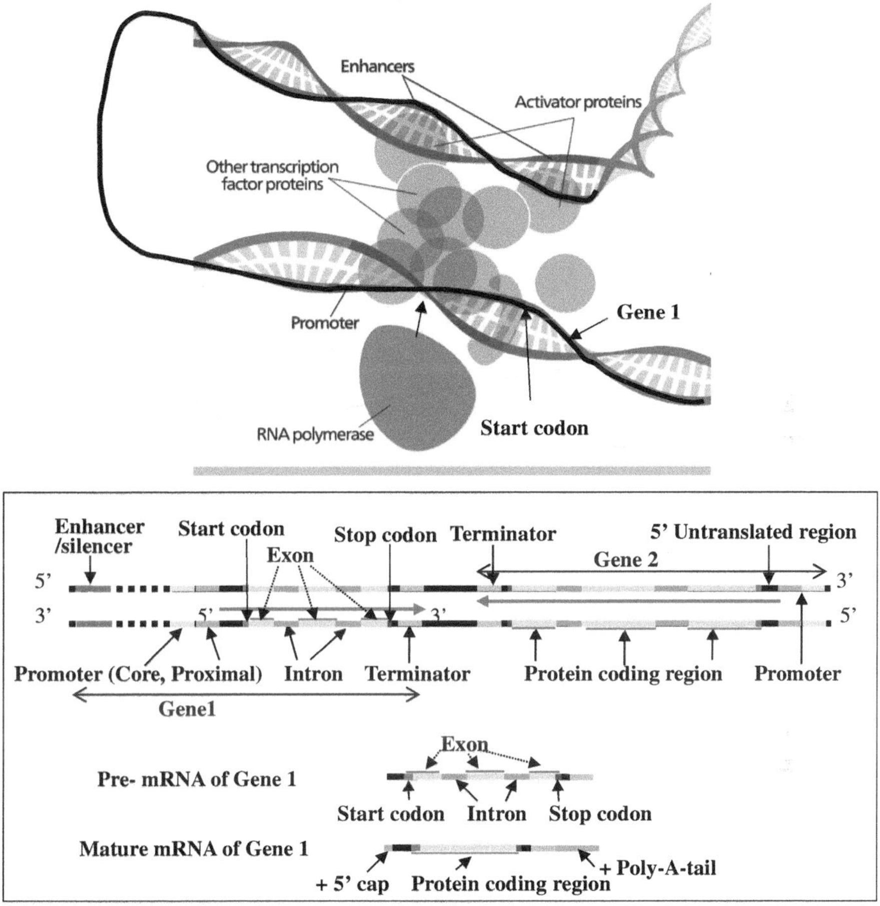

Fig. 3.7 The structure of eukaryotic protein-coding genes. Although DNA is a double-stranded molecule, only one of the strands encodes information that the RNA polymerase reads to produce mRNA. The promoter specifies where mRNA synthesis starts, and which DNA strand serves as the template strand. RNA polymerase binds the proximal promoter region aided by the transcription factor proteins bound on the core part of the promoter. To synthesize RNA in the 5′ to 3′ direction going from left to right, RNA polymerase reads the bottom strand of DNA in the 3′ to 5′ direction. To synthesize RNA in the 5′ to 3′ direction going from right to left, RNA polymerase reads the top strand in the 3′ to 5′ direction. 5′ untranslated region prior to the start codon is involved in ribosome binding sequence, and 3′ untranslated region after the stop codon is involved in terminating the transcription of the gene.

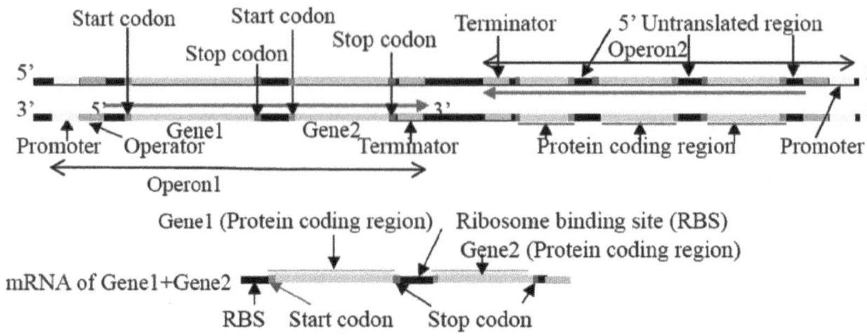

Fig. 3.8 The structure of polycistronic prokaryotic protein-coding genes. Genes coding for functionally related proteins are under the control of one promoter. Binding of the polymerase to a promoter is usually repressed as a repressor is binding to an operator for the control of the expression of a gene. As soon as an inducer binds to the repressor, the repressor is detached from the operator, and it is ready for the polymerase to bind and start transcription. An inducer is usually a substrate of the enzyme that a gene of the operon encodes. An operon has a pair of a promoter and an operator, multiple pairs of an untranslated region and a protein coded region, and a prokaryotic terminator. Each protein coded region has its own ribosome binding site (RBS) in the untranslated region so that ribosomes can simultaneously translate the coding regions on the same mRNA.

operator, and then it is ready for RNA polymerase to bind to the promoter and start transcription (Fig. 3.8). In *eukaryotes*, transcription factor proteins help position RNA polymerase and bind it to the promoter. Transcription involves transferring the complementary base sequence of the DNA to the RNA molecule. RNA polymerase reads the DNA template and synthesizes RNA in $5'$ to $3'$ direction. Genetic information is now transferred to the RNA which encodes a protein. A specific sequence of three consecutive nucleotides called a codon in protein coding regions of the transcribed RNA specifies the amino acid and the start or stop of the protein synthesis in a ribosome. *The start codon on RNA is AUG and a stop codon on RNA is either UAA or UAG or UGA*. Each codon is complementary to the corresponding triplet in the DNA from which it was transcribed.

When the polymerase reaches a special base sequence, transcription terminates. At the $3'$ end after a stop codon, there is a poly-A sequence such as AAUAAA called a terminator. This sequence marks the endpoint for transcription. It signals for RNA cleavage enzyme to cut the RNA from the transcription complex and release the RNA polymerase. In *eukaryotes*, genes contain noncoding base sequences called *introns*, and the coding regions called *exons*. When pre-mRNA is newly formed, its introns are removed, and exons are joined in the process called *splicing*. Following the cutting,

3′ poly-A tail (a 100–300 sequence of adenine nucleotides) is added to 3′ end of the pre-mRNA. Furthermore, a modified GTP cap consists of a guanine nucleotide that is connected to the 5′ end of the pre-mRNA via an unusual 5′ to 5′ triphosphate linkage (Fig. 3.7). This tail and the G cap are important because they help mRNA leave the nucleus via a nuclear pore and protects the mRNA from degradation. After bases are added at both ends, then it leaves the nucleus through a nuclear pore. The 5′ untranslated region prior to the start codon (AUG) on mRNA binds to the ribosome for translation and facilitates the transcribed protein to be functional.

On the other hand, in *prokaryotes*, genes contain no introns, and so no splicing is needed. Prokaryotic genes are grouped into a polycistronic operon. An operon has a pair of a promoter and an operator, multiple pairs of a ribosome binding site (an untranslated region) and a protein coded region sandwiched between the start codon and a stop codon, and a prokaryotic terminator. Each protein coded region has its own ribosome binding site (RBS) in the untranslated region so that ribosomes can simultaneously translate the coding regions on the same mRNA (Fig. 3.8). As the concentration of key metabolites changes the structure of the untranslated region of mRNA, the rate of translation is altered via blocking or revealing important sequence regions of the ribosome binding site [7]. Since there is no nuclear envelope, no poly-A tale is added, and the translation of mRNA begins in the ribosomes near the chromosome without processing the RNA.

3.8.3 Translation

The ribosomal RNA in the small subunit of a ribosome scans mRNA in the 5′ to 3′ direction. Translation initiation in *prokaryotes* usually requires interaction between 16S rRNA in the small subunit and an mRNA sequence upstream of the start codon, i.e., the ribosome binding site sequence called Shine-Dalgarno sequence. The small unit of a ribosome in *Escherichia coli* incorporates 16S rRNA which has the 3′-terminal sequence of 3′A-C-C-U-C-C-U-U-A_{OH}5′ which contains a complimentary sequence of 3′CCUCC5′ to a part of 5′UAAGGAGGU3 ′ known as Shine-Dalgarno sequence in the ribosome binding site of prokaryotic mRNA, and the 3′U-U-A_{OH}5′ is involved in the termination of mRNA [8]. As soon as 16S rRNA recognizes and binds to the sequence of 5′GGAGG3′ in mRNA, translation of mRNA into an amino acid sequence starts at the first AUG (start codon) and stops at UAA, UAG, or UGA (stop codon). AUG, UCC, and UUU are translated to methionine, serine, and phenylalanine, respectively, as shown in Table. 3.1. It is now clear that the Shine-Dalgarno sequence is important for the identification of the translation initiation site on the mRNA which can be recognized by the ribosome, and the spacing between the Shine-Dalgarno and the start codon strongly affects translational efficiency [9]. When the 16S rRNA scans a sequence of three bases (codon) in mRNA which encodes a protein, it calls for one amino acid which is carried by the adaptor molecule of transfer RNA (tRNA). tRNA is present at tens of millions in a human cell and is the most abundant RNA in moles among all cellular RNAs. Each tRNA brings each specified amino acid dissolved in the cytosol into the large subunit of the ribosome (see Fig. 3.9). In the large subunit, the ribozyme catalyzes peptide bond formation of these amino acids.

Question 3.2 The complete sequence of a small mRNA encoding a small protein is shown below. Predict the peptide encoded by this mRNA.

mRNA = 5' AUAGAUGAACCCCCAAGCCUAGGAG 3′

Answer Met-Asn-Pro-Gln-Ala. The ribosome starts translation at the first AUG. UAG is a stop codon.

3.8.4 Gene silencing via RNA interference

Our body consists of different cell types like skin, muscle, nerve, or bone cells. They have different structures and functions but contain similar components like the nucleus which contains genomic DNA molecules that encodes the instructions for cell development and function. This information is arranged into segments of DNA called genes. Almost every cell contains identical copies of DNA, but if there are the same instructions for all cells why are there so many different types. The reason is gene silencing. Genetic information can be switched off so that during developments a cell can only read instructions that are necessary for gaining the characteristic structure and functions. In gene silencing involved in RNA interference, a cell can use short interfering RNA (siRNA) and microRNA (miRNA) which are able to turn off some genes by inactivating mRNA. In this way, translating the gene information into proteins are regulated from its development to its death. The dysregulation can have serious consequences for our body and can cause a disease such as cancer and heart disease. Exogenous siRNA is about 21–23 nucleotides in length derived from longer double-stranded RNA. Therapeutic siRNA may be delivered into cells in gene cancer therapy though exogenous siRNA is prone to degradation and has a limited ability to cross cell membranes.

Table 3.1 Protein coding region of mRNA contains a series of specific sequences of three consecutive nucleotides (codon) that specifies the particular amino acid or the start or stop of protein synthesis.

		Second base								
		U		C		A		G		Third base
First base	**U**	UUU	Phe	UCU	Ser	UAU	Tyr	UGU	Cys	**U**
		UUC	Phe	UCC	Ser	UAC	Tyr	UGC	Cys	**C**
		UUA	Leu	UCA	Ser	UAA	Stop	UGA	Stop	**A**
		UUG	Leu	UCG	Ser	UAG	Stop	UGG	Trp	**G**
	C	CUU	Leu	CCU	Pro	CAU	His	CGU	Arg	**U**
		CUC	Leu	CCC	Pro	CAC	His	CGC	Arg	**C**
		CUA	Leu	CCA	Pro	CAA	Gln	CGA	Arg	**A**
		CUG	Leu	CCG	Pro	CAG	Gln	CGG	Arg	**G**
	A	AUU	Ile	ACU	Thr	AAU	Asn	AGU	Ser	**U**
		AUC	Ile	ACC	Thr	AAC	Asn	AGC	Ser	**C**
		AUA	Ile	ACA	Thr	AAA	Lys	AGA	Arg	**A**
		AUG	Met	ACG	Thr	AAG	Lys	AGG	Arg	**G**
	G	GUU	Val	GCU	Ala	GAU	Asp	GGU	Gly	**U**
		GUC	Val	GCC	Ala	GAC	Asp	GGC	Gly	**C**
		GUA	Val	GCA	Ala	GAA	Glu	GGA	Gly	**A**
		GUG	Val	GCG	Ala	GAG	Glu	GGG	Gly	**G**

The relationship between the base sequence and the amino acid sequence constitutes the genetic code. The basic property of the code is triplet—i.e., a linear sequence of three bases in mRNA specifies one amino acid in a protein. For example, if the codon on mRNA is AUG, AUG codes for methionine and UAA codes for stop. Alanine (Ala), Arginine (Arg), Asparagine (Asn), Aspartic acid (Asp), Cysteine (Cys), Glutamic acid (Glu), Glutamine (Gln), Glycine (Gly), Histidine (His), Isoleucine (Ile), Leucine (Leu), Lysine (Lys), Methionine (Met), Phenylalanine (Phe), Proline (Pro), Serine (Ser), Threonine (Thr), Tryptophan (Trp), Tyrosine (Tyr), Valine (Val).

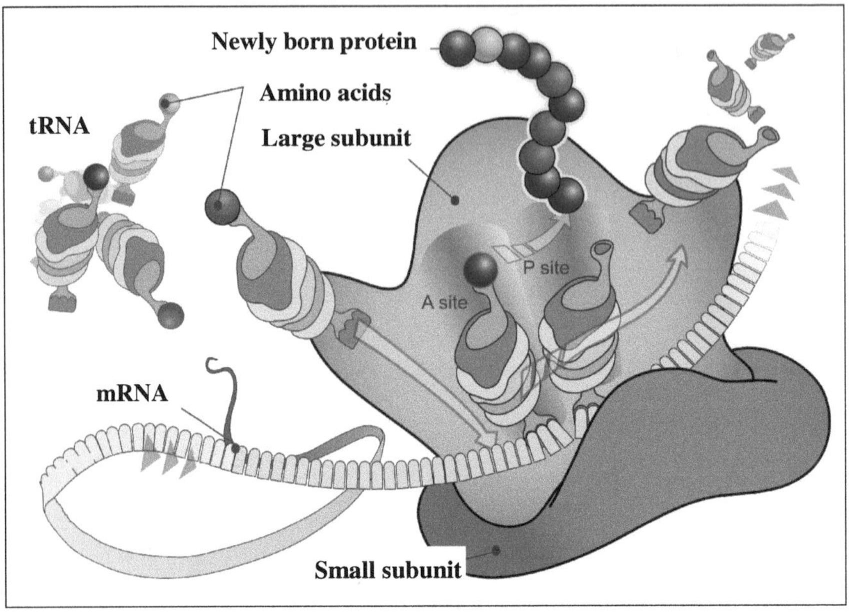

Fig. 3.9 The ribosomal RNA in the small subunit of a ribosome scans mRNA in the 5′ to 3′ direction. Translation initiation in *prokaryotes* usually requires interaction between 16S rRNA in the small subunit and an mRNA sequence upstream of the initiation codon, i.e., the ribosome binding site sequence called Shine-Dalgarno sequence. *However, many genes do not possess Shine-Dalgarno sequence.* Local absence of secondary structure (see Fig. 3.6) permits translation of mRNAs that lack ribosome-binding sites [10]. As soon as 16S rRNA recognizes and binds to the sequence of "GGAGG" in mRNA, translation of mRNA into an amino acid sequence starts at the first AUG (start codon) and stops at UAA, UAG, or UGA (stop codon). When the ribosome reads a three-base sequence on mRNA, it calls for the amino acid carried by the adaptor molecule of transfer RNA. Each tRNA brings each specified amino acid dissolved in the cytosol into a large ribosomal subunit into the large subunit. The *ribozyme* in the large subunit catalyzes peptide bond formation of these amino acids.

Genes coding for miRNA are contained in the nucleus and are transcribed by RNA polymerase II which produce premature regulatory RNA. The regulatory RNA has a typical hairpin loop structure and will become the final miRNA with regulatory function after several steps at processing. In the nucleus, the double-stranded stem is recognized by DGCR8 that is the essential cofactor of Drosha [11]. Drosha, RNase III enzyme associates with DGCR8 forming a complex which cut the RNA into a small pre-miRNA. The precursor miRNA is carried out of the nucleus through a nuclear pore by the transporter, Exportin 5. In the cytoplasm, the precursor miRNA is recognized by a large endoribonuclease, Dicer. Dicer cleaves the short double-stranded microRNA molecules about 25 nucleotides in length [12]. In the next step, argonaut protein AGO2 interacts with Dicer and binds to the microRNA and siRNA. AGO2 unwinds the double-stranded RNA, and releases one strand. The remaining strand interacts with AGO2 and some additional proteins to form the RNA-induced silencing complex, RISK. It can be guided to its target mRNA gene. Once bound, there are two ways in which RISK can inactivate the mRNA. The complex can simply cut the mRNA which will be further destroyed by the cell, or the RISK complex prevents the ribosome subunit from binding to the mRNA. In both cases, the mRNA will not be translated into a protein and the gene is silenced.

3.9 Gene cloning and expression vectors

Vector is a DNA molecule used as a vehicle to artificially carry foreign genetic materials into another cell, where it can be replicated and expressed. Bacteria have acquired additional many small circular chromosomes, i.e., plasmids to defend themselves against yeast invasions and to adapt to the environment. While chromosomes of an organism are big and contain all the essential genetic information, plasmids usually are small circular double-stranded DNA molecules (Fig. 3.10) and contain only additional genes that may be useful to the organism under certain conditions. Some of them encode for some antibiotic resistance. It has been used as a vector.

Cloning vectors: In genetic engineering, we want more copies of our gene of interest. We put it in a vector which will be taken up by bacteria and ask the bacteria to copy it. An ideal cloning vector should be a small molecular size because it is easy to handle, and its transformation into bacteria is easy.

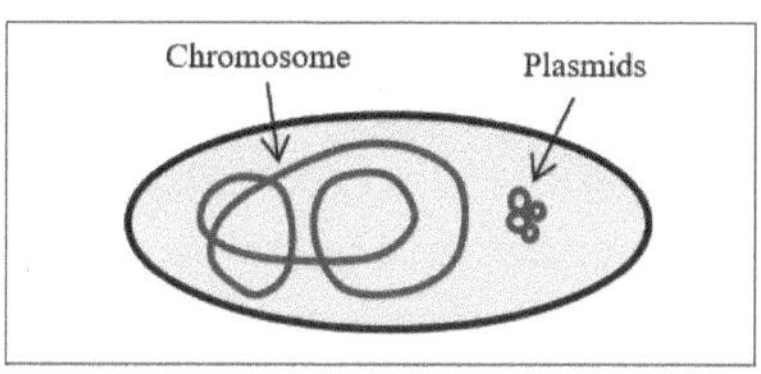

Fig. 3.10 A plasmid is a small circular DNA molecule physically separated from chromosomal DNA and can replicate independently. Plasmids are commonly found in bacteria, and some of them encode for some antibiotic resistance. It can be used as a vector.

Artificial plasmids are the most commonly used cloning vectors to drive the replication of recombinant DNA sequences within the host bacteria. Commercially available cloning vectors have the *origin of replication* to replicate independently in the host cell, and *a reporter gene* such as LacZ sandwiched between a *promoter* plus *ribosome binding site* (RBS) and a *terminator* in multiple restriction sites which has all sort of large variety of restriction sites for restriction enzymes to cut, and *selective markers* (Fig. 3.11). The gene of interest, which is cut with the same restriction enzyme, can be inserted in the cassette of the vector backbone [13]. After the gene of interest is inserted into the plasmids, they are transferred to bacteria by a process called transformation. Transformation is used to describe nonviral DNA transfer in bacteria. As the bacteria grow in the culture media, the number of the genes of interest increases.

In the culture, there are some bacteria that may take up the plasmids but there are many bacteria that cannot take the plasmid. At the end of amplification of the bacterial growth, how do we know that which bacteria have taken up the plasmids and which bacteria have not? The plasmids need to contain *selectable markers* which are usually antibiotic resistance genes conferring on "the bacteria to which the vector has been transferred" an ability to survive and proliferate in the selective growth medium containing the antibiotics. A large number of the bacteria that have not taken up the plasmid would die in the selective growth medium containing the antibiotics, and only some bacteria taking up the plasmids to which the gene of interest would be inserted are set to survive. In this way, the antibiotics act as a filter to select only the bacteria containing the plasmids.

The vectors are digested and ligated with the gene of interest at the restriction site in a reporter gene, but some of the plasmids get self-ligated and do not have the gene of interest. We need to differentiate between them whether the bacteria have taken up the plasmids with the gene of interest or they have only taken up the self-ligated plasmids. *A reporter gene* in multiple restriction sites facilitates a selection of "the host bacteria having taken up the recombinant vectors" from "the bacteria having taken up self-sealed intact ones." The reporter gene works as a screenable marker so that if it is not disrupted, expression of the enzyme can catalyze for production of a colored product whereas the bacteria having taken up the recombinant plasmids that have disrupted the reporter gene cannot express the enzyme to produce the colored product. Failing the expression can distinguish "some bacteria having taken up the recombinant plasmids" from "other bacteria having taken up self-sealed intact plasmids" (Fig. 3.11).

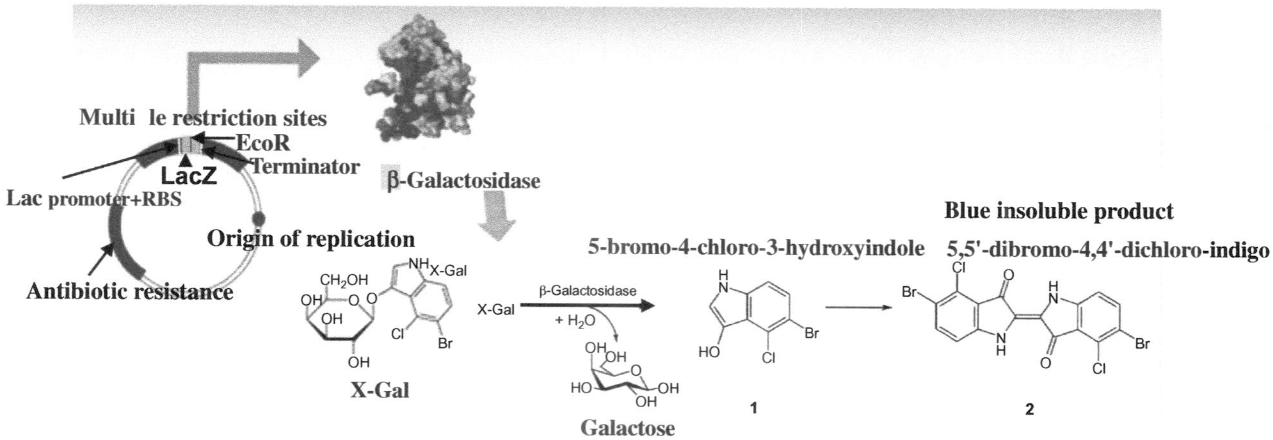

Fig. 3.11 A screenable marker in a multiple cloning site facilitates a selection of the host bacteria having taken up the recombinant vectors from the bacteria having taken up self-sealed intact ones. LacZ works as a reporter gene so that in the bacteria with the self-sealed plasmids, β-galactosidase makes a colored product when cultured in the medium containing X-Gal, and in the bacteria with the recombinant plasmids, no expression of β-galactosidase makes no colored products. Failing the expression of the enzyme can differentiate "some bacteria having taken up the recombinant plasmids" from "some bacteria having taken up self-sealed intact plasmids." The X-gal is hydrolyzed by the β-galactosidase enzyme which cleaves the β-glycosidic bond in D-lactose. X-gal, when cleaved by β-galactosidase, yields galactose and 5-bromo-4-chloro-3-hydroxyindole. The latter then spontaneously dimerizes and is oxidized into 5,5′-dibromo-4,4′-dichloro-indigo, a blue product that is insoluble. X-gal itself is colorless, so the presence of blue-colored products may therefore be used as a test for the presence of active β-galactosidase.

However, the size of a foreign DNA that can be inserted into a plasmid is only about 10,000 base pairs due to constraints on the plasmid replication limit. Moreover, another weak point of the plasmid vector is that they are hard to get through the host cell membrane since plasmids are large negatively charged polyatomic molecules. We needed an alternative vector to satisfy our requirements to carry a larger size of foreign genetic materials. Bacteriophage λ that infects *E. coli* can carry a DNA of about 45,000 base pairs. Even after eliminating the genes that cause the host cell to die and replacing them with a new DNA of our interest, the treated virus can still attach to a host cell and inject its recombinant virus DNA.

Bacteria containing the plasmid can be grown in large amounts and harvested. The plasmid of interest can be isolated from the bacteria using various methods of plasmid preparation. Many methods have been developed to purify the plasmid DNA from bacteria. These methods invariably involve three steps:

- Growth of the bacterial culture

All plasmid vectors in common use encode one or more antibiotic resistance genes as selectable markers. Bacteria that have been successfully transformed can only survive and multiply in the selective growth liquid medium containing the antibiotics such as ampicillin in the labeled microfuge tube, and thus only colonies representing successful transformations are expected to grow.

- Harvesting and lysis of the bacteria

The bacterial cells are harvested from the culture by centrifugation. A small pellet at the bottom of the tube is made up of the bacterial cells containing the plasmid DNA of interest. Remove all traces of the growth medium by pouring out wash buffer and discard the supernatant. Then it is resuspended in a resuspension buffer containing RNases to degrade RNA. Then the lysis buffer containing sodium hydroxide and sodium dodecyl sulfate is added to the suspension, but do not allow the lysis to proceed for more than 5 min in order to avoid shearing of the chromosomal DNA into small fragments, otherwise which will contaminate the plasmid DNA. Sodium dodecyl sulfate dissociates the lipid component of the cell. Contents of the cell including the cell's chromosomes and plasmid DNA spill out into the solution along with the dissociated lipids and proteins. When the bacterial cells are lysed under alkaline conditions (pH 12.0–12.5), chromosomal DNA, plasmid DNA, and proteins are denatured and dissolve in the solution.

- Purification of plasmid DNA

There exist multiple methods of nucleic acid purification. One of the methods is based on the addition of potassium acetate and acetic acid-containing neutralization buffer which returns to the pH to neutral allowing DNA strands to renature. The long single strands of chromosomal DNA get tangled with other cellular debris such as proteins and lipids to form an insoluble precipitate. Meanwhile, the smaller and intertwined plasmid DNA strands rehydrate and remain in the solution. The tube is centrifuged to separate the solid precipitate which will be discarded from the supernatant. The supernatant containing the plasmid DNA is passed through a spin column containing a silica membrane. The plasmid DNA binds to the silica membrane while the rest of the supernatant passes through the column and is discarded. The column is then washed with wash buffer which contains 70% ethanol to remove any salt contaminants. Finally, the elution buffer is added to release the plasmid DNA from the silica membrane.

Example 3.1 The artificial plasmid has an antibiotic resistance gene for ampicillin as a selectable marker, and LacZ as a reporter gene in the multiple cloning sites along with its promoter, ribosome binding site sequence (RBS), and terminator. The plasmids and a large amount of our gene of interest are incubated with a large amount of the restriction enzyme, *Eco*RV. The enzyme can recognize the sequence, GAT○ATC sites within the LacZ in the plasmid and the gene of interest and cut most of both at the recognition sites. By adding DNA ligase, the gene of interest is ligated to the plasmid at the restriction site in a reporter gene, but some of the plasmids get self-ligated and do not take the gene of interest. After the gene of interest is inserted into the plasmids, they are transferred to bacteria by a process called transformation. As the bacteria grow in the culture media, the number of the genes of interest increases. Host cells, *E. coli* used in the experiment are sensitive to ampicillin. The host cells that have incorporated the plasmids will show resistance to ampicillin. We need to differentiate among them whether the bacteria have taken up the plasmids containing the gene of interest, or they have only taken up the self-ligated intact plasmids or no plasmids. In order to clone the recombinant DNA, what kinds of medium should be used to differentiate the bacteria with recombinant DNA from others?

The bacteria used in the experiment are sensitive to both ampicillin and tetracycline. More than 99.9% of host cells will not take up negatively charged plasmid DNA since it is hard to go through the negatively charged surface of the cell membrane [14]. For the plasmid to get into the cell, the bacteria should get more competent to uptake the DNA by applying some calcium chloride and by heat shock, i.e., heating up then cooling rapidly to create lots of holes in the membrane. If the culture medium contains ampicillin, the bacteria failing to take up the plasmid die since they are sensitive to ampicillin. Only the bacteria having taken up the plasmids can survive in the medium. In the bacteria having taken up the self-sealed intact plasmid, the expression of β-galactosidase can make a colored product in the medium containing X-Gal. However, in the bacteria with the recombinant plasmids, disruption of LacZ cannot express the enzyme, and no colored product would be produced. Failing the expression of the enzyme can distinguish "some bacteria having taken up the recombinant plasmids" from "some bacteria having taken up self-sealed intact plasmids." The culture medium should contain ampicillin and X-Gal (see Table 3.2).

Expression vectors: To treat diabetes, we need insulin, but it is not practical to get a lot of insulin from humans. We put the gene of interest in a vector that will be taken up by bacteria and ask the bacteria to copy it and produce the protein encoded in the gene. The vector is used to introduce a specific gene into a target cell. Expression vectors are the basic tools in biotechnology to produce proteins of interest. An ideal expression vector should have the origin to replicate making copies and is designed for gene expression in the bacteria and contain regulatory sequences that act as a promoter, operator and RBS and a terminator along with a reporter gene such as LacZ in multiple restriction sites and selectable markers, i.e., antibiotic resistance genes. The expressing the gene for a protein can be controlled using an inducer so that the protein can be only produced in significant quantity, when necessary. Now expression vectors are commercially available. *E. coli* have been commonly used to produce insulin.

The human genes have both introns and exons. On the other hand, bacteria do not have introns in their genes, and they are only capable of reading a gene that has only exons and cannot deal with the eukaryotic gene. It is impossible for bacteria to produce the protein encoded in human genes. Since the gene of interest should contain only exons, we should make complementary DNA (cDNA) which has only exons from the isolated mRNA

Table 3.2 The insertion of a foreign DNA at the restriction site within LacZ gene makes the recombinant DNA which are taken by *E. coli* for cloning.

DNA taken by *E. coli*	Ampicillin sensitivity	Color of clone
No plasmids	Sensitive	[a]
Intact plasmids	Resistant	Blue
Recombinant plasmids	Resistant	White

[a]The culture medium contains tetracycline and X-Gal, and the host cells without the plasmid should die. Host cells, *E. coli* used in the experiment are sensitive to both ampicillin and tetracycline.

by means of reverse transcription polymerase chain reaction (RT-PCR) (see Section 3.11). The plasmids with cDNA are then transferred to *E. coli* to produce insulin (see Question 3.6).

One of the bacterial promoters is the artificially produced tac promoter. The tac promoter is a functional hybrid derived from sequences of the trp and the lac promoter. The tac promoter directs transcription approximately 11 and 7 times more efficiently than the derepressed parental *E. coli* lac UV5 promoter and approximately 3 and 2 times more efficiently than the trp promoter in the absence of the trp repressor [15]. The tac promoter is based on the lac operon, and it is normally regulated by the lac operator. When protein synthesis is called for, the transcription is initiated by the introduction of the inducer, isopropyl β-D-1-thiogalactopyranoside (IPTG) that binds to the lac repressor. Attachment of RNA polymerase to the promoter is repressed when the lac repressor is binding to the lac operator for control of the gene expression. As soon as the inducer, IPTG binds to the repressor, the repressor is detached from the operator, and then it is ready for RNA polymerase to attach to the promoter and starts transcription. The tac promoter is repressed by the lac repressor and can be derepressed by IPTG. It is used for foreign protein production in *E. coli*.

Mammalian expression vectors offer considerable advantages of expressing a gene for mammalian proteins that require the proper folding, posttranslational modifications, and relevant enzymatic activity. Production of membrane proteins requires chaperones for proper folding and numerous posttranslational modifications by means of glycosylation. Since in the eukaryotic cell, there is no concept of an operon, each gene needs each promoter to be transcribed. The artificially made plasmid should have a promoter like cauliflower mosaic virus promoter for each gene, multiple restriction sites, NGH polyadenylate chain sequence as the tag of the produced mRNA. The mammalian cell is artificially infected with the recombinant plasmid which may either stay in the cytosol or join the genomic DNA in the nucleus during mitosis.

Question 3.3 Transfection is the process of deliberately introducing naked or purified nucleic acids into eukaryotic cells. As the surface of a cell membrane is negatively charged, negatively charged DNA molecules will be repelled from the cells. What should be done for the cells to take up DNA?

Answer As the cell membrane is negatively charged, negatively charged DNA molecules will be repelled from the cell. For the cell to uptake the DNA effectively, the net charge should be positive. The main principle of using reagents is to give the DNA a net positive charge either by reagents

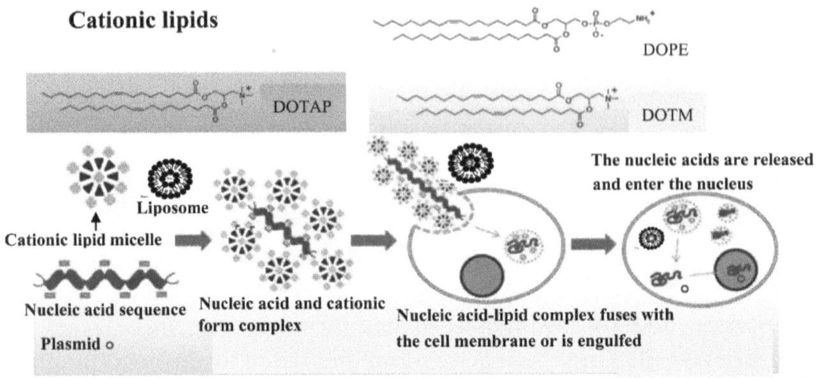

Fig. 3.12 Cationic lipid-mediated transfection: Cationic lipids are polar lipids formed of highly positively charged head groups attached to hydrophobic tails. Cationic lipids used in transfection is DOPE, DOTP, and DOTM. A lipoplexus is organized lipid molecules in the form of micelles. The net positive charge of the nucleic acid-lipid complex is effective for transfection. The lipoplexus-nucleic acid complex or liposome containing nucleic acid can be either engulfed by the cell membrane or fuse with the cell membrane. If the nucleic acid is mRNA, its effect takes place in the cytoplasm, so transfection is completed. If the nucleic acid is DNA, it needs to get into the nucleus for its effect. This happens when the cell undergoes mitosis. The nuclear membrane breaks down and some DNA gets trapped in the nuclei of the newly formed daughter cells.

binding to the DNA, or by forming complexes with DNA. The following regents have been used (1) calcium phosphate, (2) cationic lipids (Fig. 3.12), and (3) cationic polymers.

3.10 Polymerase chain reaction

In 1983, Kerry Mullis demonstrated the polymerase chain reaction (PCR) technic which made it possible to amplify genes a million times. It has been applied to the diagnosis of diseases, gene cloning, and gene identification. The PCR is based on the thermal cycle of two steps, that is, denaturing the double-stranded DNA at high 95°C and subsequent replication by elongating the DNA primers at low 70°C. PCR requires the forward and reverse primers which depends on the gene to be amplified (Fig. 3.13). First of all, we should synthesize two DNA primers complementary to the 3′ terminal end of the DNA of interest. Besides the primers, we need double-stranded DNAs to be amplified, DNA polymerase, and all deoxyribonucleoside triphosphates (dNTPs), i.e., deoxyadenosine triphosphate (dATP),

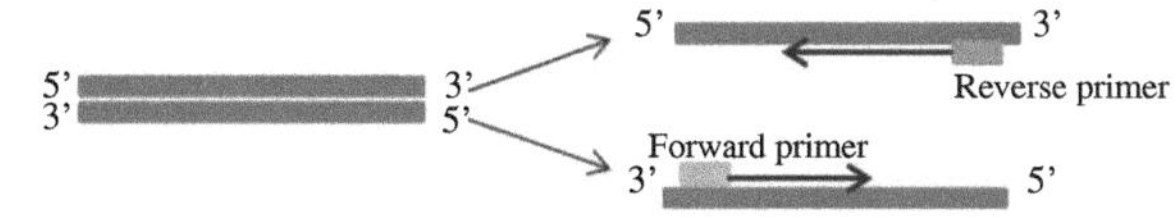

Fig. 3.13 The polymerase reaction with the forward and reverse primers at 70°C. To initiate the reaction, two kinds of short synthesized single-stranded DNAs (primers) are added in large quantities to a water solution containing double-stranded DNAs to be amplified and all deoxyribonucleoside triphosphates (dNTPs), i.e., deoxyadenosine triphosphate (dATP), deoxycytidine triphosphate (dCTP), deoxyguanosine triphosphate (dGTP), and deoxythymidine triphosphate (dTTP). These two are complementary to each 3′ end of the double-stranded DNAs.

deoxycytidine triphosphate (dCTP), deoxyguanosine triphosphate (dGTP), and deoxythymidine triphosphate (dTTP).

DNA molecules of our interest should be heated to 95°C to dissociate hydrogen bonds of double-stranded DNAs to obtain the denatured single-stranded DNA molecules. At this high temperature, ordinary DNA polymerase becomes denatured and nonfunctional. Thomas Brock and his colleagues found a unique bacterium called *Thermus aquaticus* living in the hot springs at Yellowstone. They discovered that *T. aquaticus* had all heat resistant enzymes including DNA polymerase that functioned at 95°C, and the enzyme has been used in PCR.

To initiate the reaction, two kinds of short synthesized single-stranded DNAs (primers) are added in large quantities to a water solution containing double-stranded DNAs to be amplified. These two are complementary to each 3′ end of the double-stranded DNAs. Besides those, we need the DNA replication enzyme called DNA polymerase that has been found in *T. aquaticus* and all kinds of dNTPs in abundance. The solution is then heated to 95°C to dissociate the double-stranded DNAs in a sample into the single-stranded. However, no polymerase reaction can be expected to occur at this high temperature. When the temperature is lowered to 70°C, the primers proceed to go hydrogen bonding with the target single-stranded DNA to form partial double-stranded DNA. The polymerase starts replicating by elongating the DNA primers until a complete double-stranded DNA is formed. Then the temperature is raised again to 95°C to create single-stranded DNA from the target DNA which is now doubled. The temperature is brought to be lowered again for polymerization. Polymerization needs an abundance of primers and dNTPs. The number of the DNA of interest doubles during the process every time whenever the temperature is raised and lowered. The DNA molecules are amplified by means of repeating this process.

TaqMan probe (reporter of fluorescence) consists of a fluorophor covalently attached to the 5′ end of the oligo nucleotides probe, and a quencher at the 3′ end. The sequence of nucleotides depends on the gene of interest. The close proximity of the fluorophor to the quencher prevents detection of its fluorescence. When the polymerase reaches a TaqMan probe attached to the single-stranded DNAs, its endogenous 5′ nuclease activity cleaves the probe, separating the fluorophor from the quencher to start glimmering. Thus, the light signal allows the machine to detect the duplication.

3.11 Reverse transcription polymerase chain reaction

Reverse transcription is a process to convert an RNA template to complementary DNA (cDNA) that is used in PCR amplification. The first step is the extraction of RNA from the cell or mucous fluid (see Question 3.5). The next step is the preparation of the reaction mixture for RT-PCR amplification. The RNA is transcribed to its cDNA by an enzyme called reverse transcriptase found in retrovirus. Successively, cDNA is amplified by the process called PCR (see Section 3.10). To initiate the polymerization reaction, the RNA template is added to a buffer solution containing reverse transcriptase, ribonuclease H (RNase H), all kinds of dNTPs, reverse and forward primers in large quantities, and TaqMan probe and DNA polymerase. RNase H is a family of nonsequence-specific endonuclease enzymes that catalyze the cleavage of RNA in a DNA-RNA hybrid via hydrolytic mechanism. The choice of the primers and the probe sequences depend on the target gene. The primer and probe sequences used for RNA-dependent RNA polymerase gene of SARS-CoV-2 detection were: 5′-GTGARATGGTCATGTGTGGCGG-3′ (forward), 5′-CARATGTTAAASACACTATTAGCATA-3′ (reverse) and 5′-CAGGTGGAACCTCATCAGGAGATGC-3′ (probe in 5-FAM/3′-BHQ format). The primer and probe sequences used for E gene of SARS-CoV-2 detection were: 5′-ACAGGTACGTTAATAGTTAATAGCGT-3′ (forward), 5′-ATATTGCAGCAGTACGCACACA-3′ (reverse) and 5′-ACACTAGCCATCCTTACTGCGCTTCG-3′ (probe in 5-FAM/3′-BHQ format) [16]. The temperature and duration of the reverse transcription depend on the target RNA, and reverse transcriptase. When transcribing the RNA template into its cDNA, RNase H digests the RNA in a DNA-RNA hybrid that we have just created and little fragments of the RNA template that is left. The most forward one is used as the forward primer for cDNA replication by DNA polymerase. Oligo (dT), a short deoxythymidylic acid that has a sequence of "5′-TTTTTTTT-3′," is a unique choice of the reverse primer for cDNA replication from eukaryotic mRNA since the sequence is complementary to 3′ poly-A tail.

Next, the reaction mixture is loaded in a PCR plate that is placed in a PCR machine, which is essentially a thermal cycler. Since mRNA does not last long in the cytoplasm and is present in small amounts, the newly synthesized cDNA is in small amounts and should be amplified successively using traditional PCR. Quantitative RT-PCR is considered most powerful, sensitive, and quantitative assay for the detection of RNA in tissue levels. It is frequently applied to find the specific gene for the defective protein in diseases, for using cDNA in expression vectors, or detection of COVID-19 in oropharyngeal mucus to find an infected person.

Question 3.4 How do we find a specific gene?

Answer A DNA copy of mRNA from a tissue can be made by reverse transcriptase. The complimentary DNA is different from the genomic DNA since cDNA does not contain a promoter and introns. A high degree of similarity to the cDNA is strong evidence that the region of a target genome is the protein-coding gene. And from the base sequence, we can determine the structure of the protein.

Question 3.5 How do we purify RNA from biological samples?

Answer Guanidium thiocyanate as a chaotropic solute increases the entropy of the system by interfering with intermolecular interactions caused by electrostatic forces such as hydrogen bonds and van der Waals forces, and by hydrophobic effects. Guanidium thiocyanate enhances the solvation of the hydrophobic region inside of proteins by weakening hydrophobic effects, disrupts the hydrogen binding network between water molecules, and will break phospholipid bilayers, protein-RNA complexes, its tertiary structure of proteins, and the secondary structure of RNAs.

The first step to purify RNA from biological samples is homogenization. The sample containing RNA is mixed with a lysis buffer containing mainly guanidine isothiocyanate. Guanidine isothiocyanate is a chaotropic agent that disorders water molecules and solubilizes the hydrophobic region of protein into the aqueous solution and disintegrates protein that binds to nucleic acids such as rRNA and RNA virus or denaturing enzyme that degrade RNA such as RNase. This is also directly applicable to the hydrophobic region in a lipid bilayer. If a critical concentration of the chaotropic solute is reached in the hydrophobic region of the bilayer, then membrane integrity will be compromised, and the cell will lyse. RNase inhibitors are usually present in the lysis buffer to ensure the isolation of intact RNA. Once the sample is lysed, a purification procedure is carried out by silica spin column-based method or guanidinium thiocyanate-phenol-chloroform extraction.

Silica spin column-based method: This is a solid extraction method applicable to a sample not containing DNA. Under optimal salt and PH conditions, RNA molecules are bind to a silica membrane in the spin column whereas protein and other contaminants are not retained. After the filtrate is discarded, a wash buffer containing 70% ethanol is added to the spin column containing the silica membrane to remove any remaining impurities, leaving only the RNA bound to the silica matrix. Then the final centrifugation is carried out to release the RNA from the silica membrane by forcing the elution buffer through the spin column.

Guanidium thiocyanate phenol-chloroform extraction: This is a liquid-to-liquid extraction method having a higher recovery of RNA. It relies on the phase separation by centrifugation of an aqueous sample mixed with the solution phenol-chloroform. This results in the formation of the upper aqueous layer and lower organic layer consisting of mostly phenol. Low PH around 4 ensures that only RNA preferentially remains in the aqueous phase while DNA and proteins remain in the organic layer. Isopropyl alcohol added to this aqueous layer disrupts the hydration shells neutralizing the negatively charged phosphate backbone of RNA molecules. Instead of water molecules, positively charged metal ions neutralize the negatively charged RNA. The neutralized RNA begins to aggregate in the aqueous solution. The precipitated RNA is pelleted by high-speed centrifugation and subsequently washed by 70% ethanol to remove salts and other contaminants. The purified RNA pellet is resolved in diethylpyrocarbonate, nonspecific inhibitor of RNase and stored for PCR test.

Question 3.6 How do we make insulin in bacteria?

Answer To produce human insulin in bacterial host cells, the plasmid that is chosen as a vector should have a bacterial promoter such as a tac promoter. Bacteria cannot express the genomic DNA because bacteria cannot splice the RNA. The source DNA should be cDNA which is DNA copy of mRNA for insulin, and it has no introns. Insertion of the cDNA into a plasmid and incorporation into *E. coli* cannot make the host cell produce the protein since the cDNA lacks the bacterial promoter for RNA polymerase attachment, the ribosome binding site sequence (RBS), and the bacterial terminator for transcription to end.

Commercially available expression vectors have the origin of replication to replicate independently in the host cell, and regulatory sequences that act as the promoter, operator and RBS and the terminator along with selectable makers such as antibiotic resistance genes and a reporter gene such as LacZ in multiple restriction sites which have all sort of large variety of restriction sites for restriction enzymes to cut.

When cDNA is prepared using mRNA extracted from beta cells in the pancreas and inserted into an expression vector, a bacterial clone that incorporated the vector will synthesize mRNA which will be translated into the ribosome of the host cells to produce insulin. The bacteria are cultured, and the colonies which produce insulin can be recognized by the color. The expression of a protein can be controlled using an inducer so that the protein can be only produced in significant quantity, when necessary.

3.12 Exosomal communication

Exosomes are nanoparticles that are continuously produced in most eukaryotic cells and secreted into our extracellular fluid that is our blood, saliva, or urine. Exosomes contain various molecular constituents of the cell where they are originally located, including proteins. Initially, scientists thought these exosomes were cellular garbage bags throwing out unwanted proteins in the cell. In 2007, a group led by Jan Lötvall in Sweden discovered mRNA and microRNA inside exosomes. The discovery that exosomes were transferring the genetic materials between cells suggested these particles might be involved in intercellular communication. Once into the body fluid, these tiny particles float to distant parts of the body. They carry specific cargos from the cell to very distant recipient cells in our body.

However, their small size was still a challenge for us since these tiny particles cannot be seen under a regular optic microscope, and little information was provided in electron microscopy since exosomes are too fragile to stand harsh environmental conditions like a vacuum. Atomic force microscopy made clear observation possible without damaging the fragile particles by the group at UCLA [17,18]. They could look at the vesicular shape, size, and membrane-bound structure containing many tiny globular molecules in detail.

It is becoming increasingly clear that exosomes have specialized functions and play a key role in functions including intercellular communication in the tumor microenvironment promoting cancer progression.

After the transfer of mouse exosomal RNA to human mast cells, new mouse proteins were found in the recipient cells, indicating that transferred exosomal mRNA can be translated after entering another cell. Exosomes contain both mRNA and microRNA, which can be delivered to another cell and can be functional in this new location [19]. The discovery indicates a new form of intercellular communication. Exosomes may contribute to tumor progression and metastasis by mediating cell-to-cell communication and modifying the tumor microenvironment at both local and distant sites. Certain microRNA species are found selectively in pancreatic cancer exosomes. The exosomal microRNAs have been shown to mediate intercellular communication in the tumor microenvironment and promote cancer progression [20]. The antigen-specific immunosuppression on cytotoxic T lymphocytes is mediated by myeloid-derived suppressor cells (MDSCs) which is induced by renal cancer-derived exosomes (RDEs) and promote renal tumors-growth and immune escape. HSP70, which is highly expressed in RDEs, plays a pivotal role in this process. Eliminating the function of MDSCs, or the expression of HSP70 in exosomes, or blocking the communication between RDEs and MDSCs provides future immunotherapy [21].

Living cells function by their own proteins made under the gene instructions regulated by the gene silencing mechanism. Short interfering RNA (siRNA) might be used for gene therapy since they are able to turn off some genes by inactivating mRNA. However, exogenous siRNA is prone to degradation and has a limited ability to cross cell membranes. For the cell to uptake these negatively charged nucleic acids effectively, the net charge should be positive or neutral. As exosomes naturally carry RNA between cells, these particles might be useful to deliver therapeutic siRNA to the target cells. Moreover, exosomes have their natural ability to pass through the blood-brain barrier. We could use exosomes as a natural carrier to deliver therapeutic siRNA to the distant target cells in our body in gene cancer therapy. The delivery of fluorescently labeled siRNA via exosomes to cells was confirmed using confocal microscopy and flow cytometry. The exosome-delivered siRNAs were effective at causing posttranscriptional gene silencing in recipient cells [22]. The results gave additional evidence of the ability to use human exosomes as vectors in cancer therapy, including RNAi-based gene therapy.

Also, we can use them as drug delivery agents since exosomes have their natural ability to pass through the blood-brain barrier. In order for exosomes to deliver drugs with accuracy, they need to find the sites of disease in the body. To get to the destination more easily, the outside of the exosomes should be treated with targeting molecules.

3.13 Conclusions

The self-orienting property of phospholipids in an aqueous solution leads the biological membrane into closed spherical structures and reseals it by itself when torn out. The columnar transmembrane proteins make up about half of the cell membrane and are responsible for the specialized membrane functions. The functions of the cells are regulated via signaling molecules like neurotransmitters or hormones. Living cells function by their own proteins made under the gene instructions regulated by the gene silencing mechanism. Short interfering RNA (siRNA) could be used for gene therapy since they are able to turn off some genes by inactivating mRNA. However, exogenous siRNA is prone to degradation and has a limited ability to cross cell membranes. For the cell to uptake these negatively charged nucleic acids effectively, the net charge should be positive or neutral. As exosomes naturally carry RNA between cells, these particles may be useful to deliver therapeutic siRNA to the target cells. Moreover, exosomes have their natural ability to pass through the blood-brain barrier. We can use exosomes as a natural carrier to deliver therapeutic siRNA to the distant target cells in our body in gene cancer therapy. Quantitative RT-PCR is considered most powerful, sensitive, and quantitative assay for the detection of RNA in tissue levels. It is frequently applied to find the specific gene for the defective protein in diseases, for using cDNA in expression vectors, or detection of COVID-19 in oropharyngeal mucus to find an infected person.

References

[1] T.J. McIntosh, S.A. Simon, Roles of bilayer material properties in function and distribution of membrane proteins, Annu. Rev. Biophys. Biomol. Struct. 35 (2006) 177–198.

[2] K.D. Collins, Charge density-dependent strength of hydration and biological structure, Biophys. J. 72 (1) (1997) 65–76.

[3] P. Stanley, Golgi glycosylation, Cold Spring Harb. Perspect. Biol. (2011), https://doi.org/10.1101/cshperspect.a005199.

[4] K. Barrett, S. Barman, S. Boitano, H. Brooks, Review of Medical Physiology, twenty fourth ed., McGraw-Hill, NY, 2012 (Chapter 2).

[5] E.N. Marieb, K. Hoehn, Human Anatomy & Physiology, eighth ed., Benjamin Cummings, San Francisco, CA, 2010 (Chapter 3).
[6] T.C. Westfall, D.P. Westfall, Neurotransmission: the automatic and somatic motor nervous system, in: The Pharmacological Basis of Therapeutics, eleventh ed., McGRAW-HILL, NY, 2006, pp. 137–181.
[7] T. Shafee, R. Lowe, Eukaryotic and prokaryotic gene structure, WikiJ. Med. 4 (1) (2017), https://doi.org/10.15347/wjm/2017.002.
[8] J. Shine, L. Dalgarno, The 3′-terminal sequence of *Escherichia coli* 16S ribosomal RNA: complementarity to nonsense triplets and ribosome binding sites, Proc. Natl. Acad. Sci. U. S. A. 71 (4) (1974) 1342–1346.
[9] H. Chen, M. Bjerknes, R. Kumar, E. Jay, Determination of the optimal aligned spacing between the Shine-Dalgarno sequence and the translation initiation codon of *Escherichia coli* mRNAs, Nucleic Acids Res. 22 (23) (1994) 4953–4957.
[10] L.B. Scharff, L. Childs, D. Walther, R. Bock, Local absence of secondary structure permits translation of mRNAs that lack ribosome-binding sites, PLoS Genet. 7 (6) (2011) e1002155.
[11] K.H. Yeom, Y. Lee, J. Han, M.R. Suh, V.N. Kim, Characterization of DGCR8/Pasha, the essential cofactor for Drosha in primary miRNA processing, Nucleic Acids Res. 34 (16) (2006) 4622–4629.
[12] I.J. Macrae, K. Zhou, F. Li, A. Repic, A.N. Brooks, W.Z. Cande, P.D. Adams, J.A. Doudna, Structural basis for double stranded RNA processing by Dicer, Science 311 (5758) (2006) 195–198.
[13] U.N. Streips, R.E. Yasbin, Modern Microbial Genetics, second ed., Wiley-Blackwell, 2002, p. 248.
[14] W.K. Purves, D. Sadava, G.H. Orians, H.C. Heller, Life: Science of Biology, seventh ed., W. H. Freeman and Company, Gordonsville, VA, 2003 (Chapter 16).
[15] H.A. de Boer, L.J. Comstock, M. Vasser, The tac promoter: a functional hybrid derived from the trp and lac promoters, Proc. Natl. Acad. Sci. U. S. A. 80 (1) (1983) 21–25.
[16] J. Kim, Y. Chung, H.J. Jo, N. Lee, M.S. Kim, S.H. Woo, S. Park, J.W. Kim, H.M. Kim, M. Han, Identification of coronavirus isolated from a patient in Korea with COVID-19, Osong Public Health Res. Perspect. 11 (1) (2020) 3–7.
[17] S. Sharma, H.I. Rasool, V. Palanisamy, C. Mathisen, M. Schmidt, D.T. Wong, Structural-mechanical characterization of nanoparticle exosomes in human saliva, using correlative AFM, FESEM, and force spectroscopy, ACS Nano 4 (4) (2010) 1921–1926.
[18] J.O. Olugbami, R. Damoiseaux, B. France, M.A. Gbadegesin, A.Z. Stieg, S. Sharma, O.A. Odunola, Atomic force microscopy correlates antimetastatic potentials of HepG2 cell line with its redox/energy status: effects of curcumin and *Khaya senegalensis*, J. Integr. Med. 15 (3) (2017) 214–230.
[19] H. Valadi, K. Ekstrom, A. Bossios, M. Sjostrand, J.J. Lee, J.O. Lötvall, Exosome-mediated transfer of mRNAs and microRNAs is a novel mechanism of genetic exchange between cells, Nat. Cell Biol. 9 (2007) 654–659.
[20] Y. Xu, X. Xu, A. Gin, J.D. Nshimiyimana, B.H. Mooers, M. Caputi, B.N. Hannafon, W. Ding, SRSF1 regulates exosome microRNA enrichment in human cancer cells, Cell Commun. Signal 18 (2020) 130, https://doi.org/10.1186/s12964-020-00615-9.
[21] Y. Gao, H. Xu, N. Li, H. Wang, L. Ma, S. Chen, J. Liu, Y. Zheng, Y. Zhang, Renal cancer-derived exosomes induce tumor immune tolerance by MDSCs-mediated antigen-specific immunosuppression, Cell Commun. Signal 18 (2020) 106, https://doi.org/10.1186/s12964-020-00611-z.
[22] T.A. Shtam, R.A. Kovalev, E.Y. Varfolomeeva, E.M. Makarov, Y.V. Kil, M.V. Filatov, Exosomes are natural carriers of exogenous siRNA to human cells in vitro, Cell Commun. Signal 11 (2013) 88. http://www.biosignaling.com/content/11/1/88.

CHAPTER 4

Redox reactions and ATP synthesis in the cell

4.1 Introduction

The orbital electrons neutralize the positively charged nucleus, preventing repulsion between different atoms. The outer shell electron is often referred to as the valence electron and determines the chemical properties of the element depending on the minimum energy required to remove the outer shell electron from the atom. When atoms collide to form bonding, the outermost electrons of different atoms interact constructively. If the ability of an atom to attract an electron toward itself from another atom is different, the resultant bond between them has an ionic character. The polar molecule like water is important because it will create the electric field and induce the instantaneous localization of electrons on the neighbor atom. A hydrogen bond is an electrostatic interaction between a partially positive hydrogen atom in a polar bond and a strong hydrogen acceptor of the other that is a small highly electronegative atom with a lone pair of electrons available for bonding. Nitrogen, oxygen, and fluorine are the hydrogen acceptor atoms. The redox reactions are characterized by the shift of electrons which occur between atoms of different electronegativities, and the resulting product is ionic or polar.

4.2 Shielding effects

Ionization energy (*IE*) is the *minimum* energy required to remove an outer shell electron from an atom, which is affected by the distance from the electron to the nucleus and the shielding effect that we are going to discuss. The difference of *IE* between the different elements is a prerequisite for oxidation-reduction (redox) reactions to occur through give and take of their valence electrons.

We will examine the shielding effect of the orbital electron of He atom by calculating *IE* for two extreme cases when two orbital electrons take extreme locations.

Modeling Electrochemical Dynamics and Signaling Mechanisms in Excitable Cells with Pathological Case Studies
https://doi.org/10.1016/B978-0-323-98803-2.00006-7

Extreme case #1: If electron #2 is extremely close to the nucleus and electron #1 is remarkably far from the nucleus, electron #1 experiences a minimum attractive force exerted by $Z_{eff} = +1$ instead of +2 of He atom. Because electron #2 shields electron #1 from a positive charge of one proton in the He nucleus, the orbital energy of electron #1 ($E_{e\#1}$) is considered equivalent to the orbital energy of one electron in H atom (Fig. 4.1). Z_{eff} denotes the *effective atomic number*, *n* denotes the principal quantum number, and $K = 2.18 \times 10^{-18}$ J.

$$E_{e\#1} = -IE_{e\#1} = -\frac{(Z_{eff})^2 K}{n^2} = -\frac{(+1)^2}{1^2} K = -2.18 \times 10^{-18}\,\mathrm{J}$$

Extreme case #2: If electron #1 is close to the nucleus and electron #2 is extremely far from the nucleus, electron #1 experiences a maximum force exerted by $Z_{eff} = +2$ instead of +1 because of no shielding effect.

$$E_{e\#1} = -IE_{e\#1} = -\frac{(Z_{eff})^2 K}{n^2} = -\frac{(+2)^2}{1^2} K = -8.72 \times 10^{-18}\,\mathrm{J}$$

Extreme case #1 indicates electron #1 experiences 100% shielding effect: $Z_{eff} = +1$. $IE_{e\#1} = 2.18 \times 10^{-18}$ J.

Extreme case #2 indicates electron #2 experiences no shielding effect: $Z_{eff} = +2$. $IE_{e\#1} = 8.72 \times 10^{-18}$ J.

Experimentally determined IE_{He}, 3.94×10^{-18} J is between 2.18×10^{-18} J and 8.72×10^{-18} J, and the actual shielding effect is between 0% and 100%, and closer to 100%.

Z_{eff} $(n = 1)$ can be calculated from the experimentally obtained *IE* of He.

$$Z_{eff} = \left(n^2 \frac{IE}{K}\right)^{1/2} = \left(\frac{3.94 \times 10^{-18}}{2.18 \times 10^{-18}}\right)^{1/2} = 1.34$$

The outermost electron of He atom feels the shielding effect which is found to be 67%. It is significantly large.

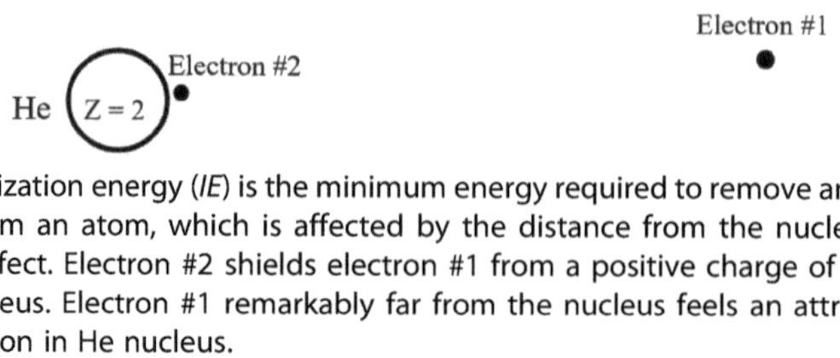

Fig. 4.1 Ionization energy (*IE*) is the minimum energy required to remove an outer shell electron from an atom, which is affected by the distance from the nucleus and the shielding effect. Electron #2 shields electron #1 from a positive charge of a proton in the He nucleus. Electron #1 remarkably far from the nucleus feels an attractive force by one proton in He nucleus.

Since the three-dimensional distribution of the *s*-electrons around the nucleus is spherical and closer to the nucleus while that of the valence *p*-electrons is shaped like a doughnut and the *p* orbitals, p_x, p_y, p_z are orthogonal to each other. The valence *p*-electrons are thought to be shielded by the *s*-electrons but less shielded by the other valence *p*-electrons, and then the *p*-electrons are pulled more by the nucleus as the number of the *p*-electrons increases. Four electrons of *s*-orbitals of fluorine atom shield electrons of *p*-orbital from the positive charge of nine protons in the fluorine nucleus. The approximate minimum energy required to remove an outermost electron from a fluorine atom is

$$IE_{2p} = \frac{(Z_{\text{eff}})^2 K}{n^2} = \frac{(9-4)^2}{2^2} K = 13.6 \times 10^{-18}\,\text{J}$$

where $Z_{\text{eff}} = Z - \sigma$. Z is the atomic number, and σ denotes the number of electrons that have the shielding effect. In this case, $\sigma = 4$.

The minimum energy required to remove an electron from a hydrogen atom is

$$IE_{1s} = \frac{(Z_{\text{eff}})^2 K}{n^2} = \frac{(+1)^2}{1^2} K = 2.18 \times 10^{-18}\,\text{J}$$

The fluorine nucleus more strongly attracts the orbital electron of hydrogen toward itself when they bump. If a hydrogen atom collides with a fluorine atom, the electron of the hydrogen orbital will be knocked out and pulled by the fluorine nucleus. The fluorine atom finally becomes more stable because of the achievement of a stable rare gas configuration.

4.3 Electronegativity

Ionization energy abbreviated to *IE* is the minimum energy required to remove the most loosely bound outermost electron of an isolated neutral gaseous atom or molecule from the orbital. From left to right within each row in the periodic table, *IE* generally increases since the outermost *p*-electron is pulled more to the nucleus along with the increased number of its protons in the nucleus. From top to bottom in each column, *IE* generally gets smaller since the outer shell electrons are farther away from the nucleus.

A photon can knock an orbital electron out of the atom if its energy is sufficient. Photoelectron spectroscopy uses X-ray energy for knocking out orbital electrons in a free atom and displays the spectrum of the ejected electrons. Ionization energy can be calculated by subtracting the kinetic energy of the ejected electron from the incident energy of the X-ray. The ionization energy is equivalent to the work function for the photoelectric effect. You may recall the work function is the energy required to eject an electron from the surface of a metal.

$$IE = E_{\text{incident}} - KE.$$

The incident energy of X-ray is $\frac{hc}{\lambda}$ where h is Plank's constant, λ is the wavelength of the incident beam, and c is the speed of light in a vacuum.

The kinetic energy of the ejected electron is calculated from the momentum (p) of the ejected electron.

$p = h/\lambda$, where λ is the wavelength of the ejected electron measured by spectroscopy.

$$KE = \frac{p^2}{2m_e} \tag{4.1}$$

where m_e is electron rest mass.

If the X-ray photoelectron spectroscopy displays an emission spectrum with four distinct kinetic energies, what are the elements which produce this spectrum? The spectrum lines originate from 1*s*, 2*s*, 2*p*, 3*s*. The elements should be Na or Mg.

Electron affinity (*EA*) is the ability of an atom to gain electrons. Chlorine tends to gain one electron from a different element on the left of a row in the periodic table and becomes Cl^- which is more stable than Cl. Linus Pauling first proposed electronegativity in 1932. Electronegativity (χ) is a measure of the ability of an element to attract electrons toward itself from another element. Atom's electronegativity is affected by both its atomic number and the distance of its valence electrons from the positively charged nucleus. Mulliken later proposed electronegativity as the average of *IE* and *EA* [1].

$$\chi = \frac{1}{2}(IE + EA) \tag{4.2}$$

The Mulliken electronegativity can only be calculated for the element if the electron affinity is known besides ionization energy.

After Mulliken, Leland C. Allen has proposed that electronegativity is related to the average ionization energy of the valence electrons in a free atom [2]. Since the valence *p*-electrons are shielded by the valence *s*-electrons but less shielded by the other valence *p*-electrons, they are pulled more by the nucleus if the number of the *p*-electrons increases. The number of p-electrons contributes a considerably important factor on electronegativity.

$$\chi = \frac{n_s IE_s + n_p IE_p}{n_s + n_p} \tag{4.3}$$

where IE_s and IE_p are the ionization energies of s- and p-electrons in the outer shell, respectively, and $n_{s,\ p}$ are the number of s- and p-electrons in the valence shell. It is to apply a scaling factor, 1.75×10^{-3} for energy measured in kJ/mol to give values that are numerically equivalent to Pauling's electronegativity. Electronegativities of some elements were calculated according to Eq. (4.3) using results of the emission spectrum displayed by the X-ray photoelectron spectroscopy.

$$\mathrm{H}\,(\mathrm{Z}=1,\ n_{1s}=1,\ IE_{1s}=1312\,\mathrm{kJ/mol}),\ \chi = 2.18 \times 10^{-21} \times 6.02 \times 10^{23} \times 1.75 \times 10^{-3} = 2.29$$

$$\mathrm{Na}\,(Z=11,\ n_{3s}=1,\ IE_{3s}=496\,\mathrm{kJ/mol}),\ \chi = 496 \times 1.75 \times 10^{-3} = 0.89$$

$$\mathrm{Mg}\,(Z=12,\ n_{3s}=2,\ IE_{3s}=738\,\mathrm{kJ/mol}),\ \chi = 738 \times 1.75 \times 10^{-3} = 1.29$$

$$\mathrm{Al}\,\left(Z=13,\ n_{3s}=2,\ n_{3p}=1,\ IE_s=1816\,\mathrm{kJ/mol},\ IE_p=577\right),$$

$$\chi = \frac{(2 \times 1816 + 577) \times 1.75 \times 10^{-3}}{3} = 2.45$$

$$\mathrm{Si}\,\left(Z=14,\ n_{3s}=2,\ n_{3p}=2,\ IE_s=1577\,\mathrm{kJ/mol},\ IE_p=786\right),$$

$$\chi = \frac{(2 \times 1577 + 2 \times 786) \times 1.75 \times 10^{-3}}{4} = 2.07$$

$$\mathrm{P}\,\left(Z=15\ n_{3s}=2,\ n_{3p}=3,\ IE_s=1890\,\mathrm{kJ/mol},\ IE_p=1060\right),$$

$$\chi = \frac{(2 \times 1890 + 3 \times 1060) \times 1.75 \times 10^{-3}}{5} = 2.43$$

$$\mathrm{S}\,\left(Z=16,\ n_{3s}=2,\ n_{3p}=4,\ IE_s=2295\,\mathrm{kJ/mol},\ IE_p=1000\right),$$

$$\chi = \frac{(2 \times 2295 + 4 \times 1000) \times 1.75 \times 10^{-3}}{6} = 2.51$$

$$\mathrm{Cl}\,\left(Z=17,\ n_{3s}=2,\ n_{3p}=5,\ IE_s=2260\,\mathrm{kJ/mol},\ IE_p=1256\right),$$

$$\chi = \frac{(2 \times 2260 + 5 \times 1256) \times 1.75 \times 10^{-3}}{7} = 2.7$$

Ionization energies of Na, Mg, and Al are smaller than that of hydrogen due to the distance of its valence electrons from the charged nucleus and the shielding effect from the inner shell electrons.

Electronegativity is a measure of the ability of an atom to attract an electron toward itself from another atom. From left to the right within a given row in the periodic table, electronegativity (χ) generally increases. From top to the bottom in each column, electronegativity generally decreases. If an atom on the left side of the periodic table collides with an atom on the right, an atom on the left side loses one or more electrons while at the same time an atom on the right gains one or more electrons (Table 4.1). A bond between atoms with similar negativities is called a nonpolar covalent bond. The shared electrons are more likely near one atom than the other if the bond forms between different elements. The bond between different kinds of atoms with the difference of electronegativity between 0.4 and 1.7 is considered a polar covalent bond. The bond between different kinds of atoms with the difference of electronegativity over 1.7 is considered an ionic bond. The hydrogen bond is an electrostatic interaction between a hydrogen atom in a polar molecule and a hydrogen acceptor in the other polar molecule.

Table 4.1 Periodic table with electronegativity.

H																	He	
2.20																		
2	Li	Be											B	C	N	O	F	Ne
	0.98	1.57											2.04	2.55	3.04	3.44	3.98	
3	Na	Mg											Al	Si	P	S	Cl	Ar
	0.93	1.31											1.61	1.90	2.19	2.58	3.16	
4	K	Ca	Sc	Ti	V	Cr	Mn	Fe	Co	Ni	Cu	Zn	Ga	Ge	As	Se	Br	Kr
	0.82	1.00	1.36	1.54	1.63	1.66	1.55	1.83	1.88	1.91	1.90	1.65	1.81	2.01	2.18	2.55	2.96	3.00
5	Rb	Sr	Y	Zr	Nb	Mo	Tc	Ru	Rh	Pd	Ag	Cd	In	Sn	Sb	Te	I	Xe
	0.82	0.95	1.22	1.33	1.6	2.16	1.9	2.2	2.28	2.20	1.93	1.69	1.78	1.96	2.05	2.1	2.66	2.60
6	Cs	Ba	*	Hf	Ta	W	Re	Os	Ir	Pt	Au	Hg	Tl	Pb	Bi	Po	At	Rn
	0.79	0.89		1.3	1.5	2.36	1.9	2.2	2.20	2.28	2.54	2.00	1.62	1.87	2.02	2.0	2.2	2.2

Electronegativity is the ability of an atom to attract an electron toward itself from another atom. Bond between atoms with similar negativities is called nonpolar covalent bond. Bond between different kinds of atoms with the difference of electronegativity between 0.4 and 1.7 is considered as polar covalent bond. Bond between different kinds of atoms with the difference of electronegativity over 1.7 is considered as ionic bond. Hydrogen bond is an electrostatic interaction between a hydrogen atom in a polar molecule and a hydrogen acceptor in the other polar molecule. The hydrogen acceptors are small highly electronegative atoms with a lone pair of electrons such as nitrogen, oxygen, and fluorine.

The hydrogen acceptors are small highly electronegative atoms with a lone pair of electrons such as nitrogen, oxygen, and fluorine. For example, water is a polar molecule which has one oxygen atom with two lone pairs and two hydrogen atoms. It will create the electric field and induce the instantaneous localization of electrons on the neighbor atom. When two water molecules get close together, the electrostatic forces work to attract the molecules together. The oxygen atom of one water molecule will bond with two hydrogen atoms of other water molecules. These bonds are called hydrogen bonds. Hydrogen bonds are not as strong as covalent bonds, but they are strong enough to bind water molecules together and give water its unique characteristics.

4.4 Regulation of cell's function via signaling molecules

In inorganic chemistry, a ligand is an ion or molecule that binds to a central metal atom to form a coordinate complex. In biochemistry and pharmacology, a ligand means a substance that forms a complex with a biomolecule to serve as signaling molecules to elicit a biological effect in contrast to the definition in inorganic chemistry. In protein-ligand binding, a ligand is usually a small molecule that produces a signal by binding specifically to the site called a receptor on a target protein. Ligands may be hormones such as epinephrine (EPI) or neurotransmitters such as GABA, glutamate, acetylcholine (Ach), or norepinephrine (NE). In addition, ligands also include substrates of the enzyme, and second messengers such as cAMP, Ca^{2+}, inositol triphosphate (IP_3), and diacylglycerol (DG). Binding between a ligand and its receptor results from intermolecular forces such as electrostatic forces or van der Waals forces, which cause the conformational change of the target protein. The binding typically changes the three-dimensional shape of the target protein, which switches on various physiological effects depending on the target protein. The target protein may be (1) enzymes, (2) ligand-gated ion channels, or (3) G protein-coupled transmembrane proteins that bind selectively to hormones or neurotransmitters.

The binding occurs over an infinitesimal range of time. It is reversible, and the dissociation is fast. Binding affinity can be changed by solvent effects.

(1) Enzymes change the shape to hold charged or polar substrates, and the conformational change leads to catalyze the substrates. If some molecules bind to the active site of an enzyme more strongly than the substrate and the enzyme cannot cleave them, they bind at the active site of the enzyme and thus prevent the substrate from binding. A substance that structurally resembles the substrate can be a candidate of the inhibitor used as a drug.

(2) When binding to neurotransmitters or second messengers, the ligand-gated ion channels open through the conformational change of the pore protein.

(3) The binding of the ligands activate an associated G protein by exchanging the GDP (which is bound to the G protein) with GTP, and then the G protein's α subunit with GTP dissociates from the βγ subunits and further affects intracellular target enzymes such as adenylate cyclase and phospholipase C directly depending on the α subunit type. Cardiac and smooth muscle activity is modified via cyclic AMP, and IP_3 which are second messengers produced by adenylate cyclase and phospholipase C, respectively.

Selective ligands tend to bind to extremely limited kinds of receptors, whereas nonselective ligands bind to several types of receptors. This plays an important role in pharmacology, where drugs that are nonselective tend to have more adverse effects because they bind to several other receptors in addition to the one generating the desired effect.

Example 4.1 ACh, neurotransmitter, secreted from the terminal of the exciting postganglionic parasympathetic nerve binds reversibly to the muscarinic cholinergic receptors by electrostatic force induced by its positive charge on the quaternary ammonium and negative carbonyl oxygen (Fig. 4.2). The M_3 receptor is a G_q protein-coupled receptor which is abundant in smooth muscle, and the effector of α subunit of G_q ptotein is the inositol triphosphate-producing enzyme, phospholipase C. The dissociated α subunit of G_q protein are responsible for the activation of phospholipase C which catalyzes the hydrolysis of phosphatidylinositol 4,5

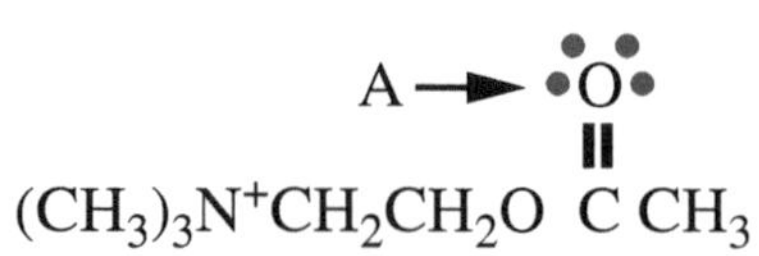

Acetylcholine

Fig. 4.2 ACh is a diffusible neurotransmitter in an aqueous solution because of a hydrophilic molecule with a positive charge on quaternary ammonium and negative carbonyl oxygen. ACh is attracted by the anionic site of the receptor of ACh-gated Na^+ channel protein or G protein-coupled receptor, or the anionic site of acetylcholinesterase and binds the site A to the activation site of the receptor or the catalytic site of the enzyme.

bisphosphate that is one of the plasma membrane phospholipids. The hydrolysis produces the second messengers, inositol triphosphate (IP_3) and diacylglycerol (DAG) which are playing an important role in signal transduction in biological cells. IP_3 is hydrophilic and diffuses in the cytosol, where it binds to its receptor on IP_3-gated Ca^{2+} channels in the endoplasmic reticulum (ER) membrane, which subsequently increases a release of Ca^{2+} from intracellular stores in the ER by opening, and the released second messenger, Ca^{2+} activates various calcium-regulated intracellular signaling systems, while hydrophobic DAG stays inside the membrane, and its van der Waals forces activates protein kinase C (PKC) in the presence of Ca^{2+} (see Section 8.5). Then the activated PKC subsequently opens L-type Ca^{2+} channels by phosphorylation. In this example, a neurotransmitter (ACh) and second messengers (Ca^{2+}, IP_3, and DAG) are ligands.

4.5 Redox reactions between molecules with an atom of different electronegativity

Oxidation was originally defined as the combination of oxygen with other elements. Oxygen has a high ability to attract two electrons toward itself from the other atom, and an electron of hydrogen tends to be pulled toward the neighbor oxygen, and one negatively charged oxygen atom attracting two electrons from hydrogen molecules and two positively charged hydrogen atoms combine to produce water. The hydrogen is oxidized, and the oxygen is reduced. The definition has been expanded further that oxidation means loss of one or more electrons from a substance, and reduction means gaining of one or more electrons to a substance. Oxidizing agents such as oxygen, nitrogen, fluorine, and NAD^+ have an atom with higher electron affinity and attract electrons from neighbor molecules that have an atom with lower electron affinity, and the oxidizing agent itself is reduced. Reducing agents such as hydrogen, vitamin C, and NADH have an atom with lower electron affinity, shift its electron to another one that has an atom with higher electron affinity, and the reducing agent itself is oxidized. The oxidation-reduction (redox) reaction may occur between molecules with the atom of different electronegativity. Between the reducing agent and the oxidizing agent, there is a difference in electric potential. Redox reactions are characterized by the shift of electrons from the reducing agent at the lower electric potential to the oxidizing agent at the higher electric potential.

4.6 Synthesis of ATP driven by electrochemical potential

Reducing agents, NADH and $FADH_2$ are by-products of sugar metabolism in mitochondria which start electron transport in the inner mitochondrial membrane to produce ATP. Without oxygen supply, the electron transfer stops, and the ATP synthesis comes to a halt. Nicotinamide adenine dinucleotide (NAD^+) and flavin adenine dinucleotide (FAD) are oxidizing agents. NAD^+ is notable for oxidation of pyruvate, isocitrate, ketoglutarate, and malate, and FAD is notable for oxidation of succinate, respectively. NAD^+ and FAD accept two electrons from the molecules produced in the process of sugar metabolism, bind to one or two protons from the matrix, respectively and become reducing agents, i.e., NADH and $FADH_2$ (Fig. 4.3). In the inner mitochondrial membrane, the four respiratory protein complexes I, II, III, and IV are embedded and involved in transferring electrons via the mitochondrial electron-transport chain to the strongest final electron acceptor, i.e., oxygen. Electrons are transferred from a molecule with low electron affinity to high electron affinity along the electric potential gradient to the final acceptor, oxygen, by means of the mitochondrial electron-transport chain. Electron-transport chain produces an electrochemical potential difference of protons between the matrix and the intermembrane space of mitochondria via redox mechanism during aerobic respiration in mitochondria. The created potential energy of protons is used to drive F_0F_1-ATP synthase to produce ATP.

When NADH passes electrons to the oxidized form of membrane soluble coenzyme Q (CoQ) in the presence of the first mitochondrial respiratory enzyme, NADH-CoQ oxidoreductase in the complex I, releasing protons into the intermembrane space, and $FADH_2$ pass electrons to the CoQ in the presence of $FADH_2$-CoQ oxidoreductase in the small complex II, respectively, the CoQ becomes the reduced form accepting the electrons. The electrons are carried by the CoQ from complexes I and II to complex III. In the presence of CoQ-cytochrome *c* oxidoreductase in the complex III, the electrons are passed to peripheral membrane protein cytchrome c, releasing protons into the intermembrane space, and then the cytochrome *c* carries the electrons from complexes III to IV. Complex IV contains cytochrome *c* oxidase which is the terminal enzyme of the electron transport chain. When the strongest electron acceptor, O_2 accepts four electrons from the cytochrome *c*, the cytochrome *c* oxidase catalizes the redox reaction between O_2 and four protons to yield two H_2O. As electrons are carried from Complexes I and II to Complex IV through a series of redox centers

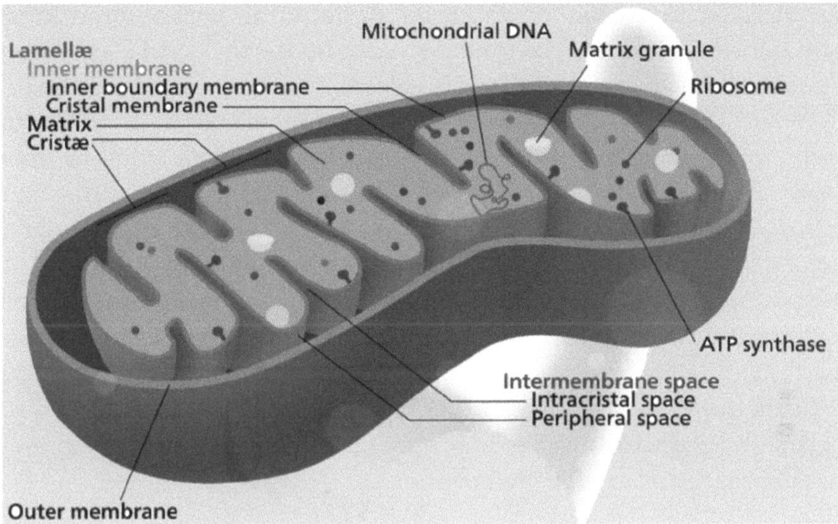

Fig. 4.3 The successive electron transport process is as follows. Oxidative decarboxylation of pyruvate and the citric acid cycle takes place in the mitochondria matrix. → NADH or $FADH_2$ are produced. → NADH and $FADH_2$ carry electrons to the respiratory enzyme complex I and complex II, respectively. → Electrons are transferred between complexes I and III by the mobile membrane soluble coenzyme, CoQ and between complexes III and IV by the peripheral membrane protein cytochrome *c*. Complex II transfers electrons from succinate to CoQ, then CoQ carries electrons from complex II to complex III. Finally, oxygen in the matrix accepts electrons from the complex IV. → While electrons are carried from Complexes I and II to Complex IV through a series of redox centers which are clusters of atoms, protons move through the complexes I, III, and IV into the intermembrane space of mitochondria from the matrix, which in turn creates the electrochemical potential difference between them. → The potential difference is strengthened as 10 protons are carried out into the intermembrane space while 4 protons in the matrix are consumed to produce 2 water molecules. The creation of the concentration gradient of a proton between inside and outside the inner mitochondria membrane drives the F_0F1-synthase to synthesize ATP.

which are clusters of atoms, protons move through the complexes I, III, and IV into the intermembrane space of mitochondria from the matrix, which in turn creates the electrochemical potential difference between them. The potential difference is strengthened as protons are carried out into the intermembrane space from the matrix while 4 protons in the matrix are consumed to produce 2 water molecules.

The concentrated protons in the intermembrane space diffuse back to the matrix, driving complex V called F_0F_1-ATP synthase in the inner mitochondria membrane to catalyze the ADP to ATP conversion by rotating the lower subunit of the synthase continuously. F_0F_1-ATP synthase is composed of two mechanical rotary motors. The lower hydrophobic part (F_0 motor) of F_0F_1 ATP-synthase in the inner membrane is driven by the proton influx along the electrochemical potential gradient from the intermembrane space to the matrix (Fig. 4.4). The forced rotation of the upper hydrophilic part (F_1 motor) converts the electrochemical potential difference of the protons into chemical energy to synthesize ATP from ADP and inorganic phosphate via the mechanical rotation of the rotary shaft connected to the F_0 motor [3]. Without sufficient oxygen, the electron transfer stops, and the electrochemical potential difference of the protons diminishes when the ATP synthesis comes to a halt, and then the upper part of the enzyme (F_1) starts rotating in the reverse direction as a proton pump using bonding energy from ATP hydrolysis.

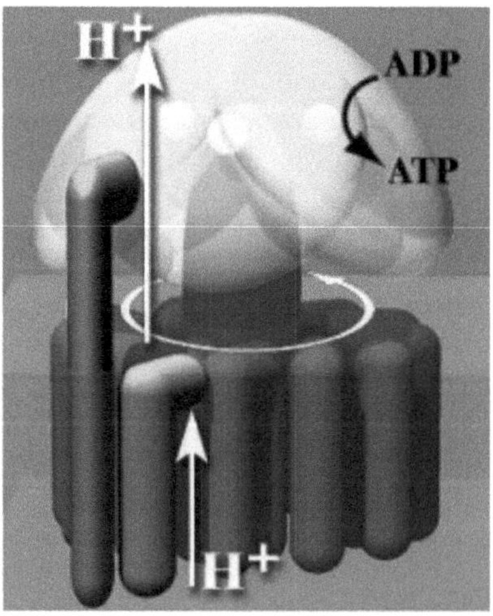

Fig. 4.4 The upper hydrophilic part of F_0F1 ATP-synthase is the F_1 motor protruded in the matrix and the lower hydrophobic part is the F_0 motor of F_0F1 ATP-synthase in the inner mitochondria membrane. The two molecular motors convert the electrochemical potential difference of the protons into the chemical energy to synthesize ATP from ADP and inorganic phosphate via the mechanical rotation of the rotary shaft. Without sufficient oxygen, the electron transfer stops, and the electrochemical potential of the proton diminishes while the ATP synthesis comes to a halt, and then the upper part of the enzyme (F_1) starts rotating in the reverse direction as a proton pump using high-energy phosphate from ATP hydrolysis.

4.7 Relationship between electric work and Gibbs free energy

Let us calculate the change in Gibbs free energy in the reversible redox reaction where transfer of electrons from the element of a molecule with low electron affinity to the element of a molecule with high electron affinity under constant temperature and pressure. The element with low electron affinity such as Zn creates low electric potential around it when it is solved in acidic solution as an anode (electron donor), releasing electrons to the connecting wire. Molecules with high electron affinity such as Cu creating high electric potential attract electrons from Zn through the connecting wire, and are used as a cathode (electron acceptor). The redox reaction keeps going if electrons accepted by Cu are able to be transferred to Cu^{2+} in the solution around the cathode.

In a process under constant temperature (T) (see Section 2.6.3),

$$dG = dH - TdS$$

If the system changes under constant external pressure (p_{ext}) (see Section 2.1),

$$dU = dH - p_{ext}dV$$

U is the potential energy and G is Gibbs free energy, respectively. When dG is negative, the process goes forward until equilibrium is achieved.

In a process under constant temperature and pressure,

$$dG = dH - TdS = dU + p_{ext}dV - TdS \tag{4.4}$$

If the process is reversible, we can replace q_{rev} with the state function, TdS (see Section 2.6.3).

$$dU = q_{rev} + w_{rev} = TdS + w_{rev}$$

If the reversible process is carried out under constant temperature and pressure, we obtain

$$dG = TdS + w_{rev} + p_{ext}dV - TdS = w_{rev} + p_{ext}dV \tag{4.5}$$

The system might do both nonexpansion work such as electrical work and expansion work.

$$w_{rev} = w_{rev,\ e} - p_{ext}dV$$

Substituting $w_{rev,\ e} - p_{ext}dV$ for w_{rev} in Eq. (4.5), we obtain

$$\Delta G_{elec} = w_{rev,e} \tag{4.6}$$

where $w_{rev,\ e}$ is the reversible electric work. If $w_{rev,\ e}$ is negative, the current generated in the redox reaction power the electric apparatus [4]. If the change in Gibbs free energy is negative, it means that the system has the ability to do electric work to the surrounding. If the change in Gibbs free energy is positive, it means that the surrounding does electric work to the system under constant temperature and pressure. Most of the electric potential energy created in the complexes, I, III, and IV in the inner mitochondrial membrane is used for pumping protons from the matrix into the intermembrane space of mitochondria. The creation of the concentration gradient of a proton between inside and outside the inner mitochondria membrane drives ATP synthase to catalyze ATP production.

4.8 Electric potential difference caused by electron transport in mitochondria

An oxidized form of a molecule that has more ability to attract electrons than the other becomes an electron acceptor, and the reduced form of a different kind of molecule becomes an electron donor. When they bump, the oxidized one becomes the reduced form, and the reduced one becomes the oxidized form. The standard potential of oxidized species increases along with an increase in its electron affinity. The standard biochemical conditions are set as 1 mol/L of reactants, 1 mol/L of products. The standard potentials at 25°C of each biological molecule are measured relative to the standard hydrogen electrode and can be referred to in the table.

The standard potential difference, $\Delta \overset{\circ}{V}$ for the redox reaction associated with electron transport in mitochondria is

$$\Delta V = \overset{\circ}{V}_{O/H_2O} - \overset{\circ}{V}_{NAD^+/NADH}$$

O_2 is a strong acceptor of electrons while NAD^+ is a weak acceptor. $\overset{\circ}{V}_{NAD^+}$ is negative which means it is a relatively weaker electron acceptor than hydrogen at standard conditions. The two electrons from NADH are not pulled by NAD^+, but are pulled by CoQ in the presence of the NADH-CoQ oxidoreductase in the complex I, and from CoQ to cytochrome *c* in the presence of the CoQ-cytochrome *c* oxidoreductase in the complex III, releasing a proton into the intermembrane space through the complexes. The final electron acceptor, an oxygen atom, attracts the electrons strongly toward itself, and then the negatively charged oxygen atom after accepting two electrons combines with two protons from the matrix, to produce water by cytochrome *c* oxidase, releasing two protons into the intermembrane space. NADH is oxidized, and oxygen is reduced.

$$NAD^+ + H^+ + 2e \rightarrow NADH, \quad V^{\circ}_{NAD^+/NADH} = -0.315\,V$$

$$\frac{1}{2}O_2 + 2e + 2H^+ \rightarrow H_2O, \quad V^{\circ}_{O/H_2O} = +0.615\,V$$

Since $\frac{1}{2}O_2$ has more ability to attract electrons than NAD^+ has, NADH donates two electrons to $\frac{1}{2}O_2$, and then $\frac{1}{2}O_2$ accepts two electrons and two protons. H_2O is a reduced form of $\frac{1}{2}O_2$.

The overall reaction is

$$NADH + \frac{1}{2}O_2 + H^+ \rightarrow H_2O + NAD^+$$

$$\Delta V^{\circ} = V^{\circ}_{O_2/H_2O} - V^{\circ}_{NAD^+/NADH}$$

$$= 0.615 - (-0.315) = 0.93\,V$$

The change in Gibbs free energy is

$$\Delta G^{\circ}_{elec} = -nF\Delta V^{\circ}$$

where Faraday's constant is 96,485 C per mole of electrons and n is the number of moles of electron transferred from 1 mol of NADH. $1\,V = 1\,J\,C^{-1}$.

For 1 mol of NADH oxidation

$$\Delta G^{\circ}_{elec} = -nF\Delta V^{\circ}$$

$$= -2\frac{\text{mol of } e}{\text{mol of reactant}} \times 96{,}485\frac{C}{\text{mol of } e} \times 0.93\,J\,C^{-1}$$

$$= -179\,kJ\,mol^{-1}$$

The oxidation of 1 mol of NADH by oxygen under the standard biochemical conditions releases 179 kJ of free energy [5]. Most of this energy will be used to synthesize ATP from ADP in mitochondria. The ability to do work to the surrounding attributed to the actual potential difference (ΔV) decreases as the amount of oxygen and NADH decreases, while the ability to do work to the surrounding increases as the amount of oxygen and NADH increases.

4.9 ATP consumption in the cell

Glucose is the most important energy source for living cells. Electrons released during glucose catabolism are used to take protons from the matrix into the intermembrane space and used for the synthesis of adenosine triphosphate (ATP) in mitochondria. ATP provides energy that is immediately usable by all body cells. ATP has three phosphate groups, and each phosphate group has a negative charge and repels each other. So, ATP is an extremely unstable reactive molecule with many nonbonding electrons and has two high-energy phosphate bonds which are necessary to overcome the repulsion as shown in Fig. 4.5.

When two polar water molecules attack and detach the terminal phosphate group of ATP, the hydrolysis of ATP produces a more stable molecule called adenosine diphosphate (ADP) while simultaneously releasing high bond energy attached to the phosphate. This phosphate with the high bond energy is used in the phosphorylation as seen in glucose-6-phosphate production from glucose and Na^+-K^+ ATPase activation. The unfavorable phosphorylation reactions can proceed forward when coupled with ATP hydrolysis. The phosphorylated molecule becomes more reactive. In this way, high bond energy of ATP is used when the cell needs the energy to proceed with the physiological effect.

The polarizability of P is much larger than the polarizability of O. Hydrolysis of the terminal high energy phosphate bonding to ATP is a good example of the induction of instantaneous polarization of electron

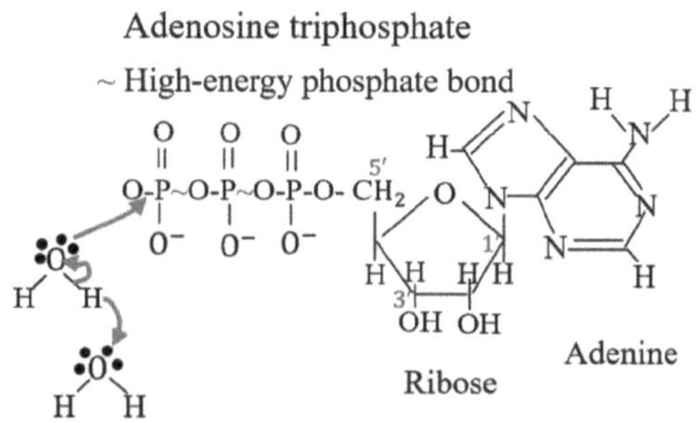

$$ATP + 2H_2O\ (l) \longrightarrow ADP + HPO_4^{-2}\ (aq) + H_3O^+\ (aq) + 30.5\ kJ$$

Fig. 4.5 Since the polarizability increases with increasing the number of valence electrons that are less tightly bound, the polarizability of P is larger than the polarizability of O. Hydrolysis of the terminal high energy bond of ATP is a good example of the induction of instantaneous polarization of electron on the P atom by two attacking water molecules, leading to detach a high energy phosphate molecule, form ADP. This high-energy phosphate is used in the phosphorylation by the kinase as seen in Glucose-6-phosphate production from glucose and Na^+-K^+ ATPase activation. The unfavorable phosphorylation reactions can proceed forward when coupled with ATP hydrolysis. The phosphorylated molecule becomes more reactive. In this way, high bond energy of ATP is used when the cell needs the energy to do physiological work.

on the P atom caused by the attack of two water molecules, leading to detach a high-energy phosphate molecule and form ADP as shown in Fig. 4.5. The released high-energy phosphate is used for the active transport, contraction of muscle fiber, or ATP-coupled reaction of a sugar.

Examples of the cellular work driven by the high-energy phosphate detached from ATP hydrolysis are as follows.

The active transporter such as Na^+-K^+ ATPase and plasma Ca^{2+} ATPase can transport the ions across the cell membrane against electrochemical potential.

The cyclic contraction of cardiac muscle is made possible by the activated myosin ATPase in the presence of Ca^{2+}. The high-energy phosphate detached from ATP in myosin causes a recocking of the myosin head to be ready for rebinding to actin (see Section 8.4).

Phosphorylation of glucose by the hexose kinase proceeds glycolysis for producing energy.

Highly reactive ATP is known as one of the units of RNA, and deoxy-ATP is known as one of the units of DNA.

4.10 ATP deficiency in the heart

Myocardial ischemia occurs when blood flow to the heart is reduced, preventing the cardiac muscle from receiving oxygen. The reduced blood flow is usually a result of a partial or complete blockage of coronary arteries. Without sufficient oxygen, the electron transfer stops in mitochondria, and the ATP synthesis comes to a halt, and then F_0F_1-ATP synthase works in the reverse direction as proton pump using bonding energy from ATP hydrolysis, and then ATP-deficiency may disturb for Na^+-K^+ ATPases to pump Na^+ out of the cytosol. Myocardial ischemia is responsible for angina and, less commonly, shortness of breath secondary to ischemic left ventricular dysfunction along with cardiac arrhythmias [6]. The acidification and cytosolic Ca^{2+} overload in the cardiomyocytes have been reported in myocardial ischemia [7]. Ca^{2+} overload also can be seen in toxic concentrations of digitalis due to an increase in cytosolic Na^+ level through the inhibition of the Na^+-K^+ pump [8].

First, during ischemia, anaerobic metabolism prevails, which produces a decrease in cell pH. To buffer this accumulation of hydrogen ions, the Na^+/H^+ exchanger excretes excess hydrogen ions, which produces a large influx of sodium ions [9]. Additionally, inactivation of Na^+-K^+ ATPase may increase more cytosolic Na^+. The high cytosolic Na^+ may reverse the action of the Na^+-Ca^{2+} exchanger and then cause an increase in the cytosolic Ca^{2+}.

Secondly, ATP deficiency due to a shortage of oxygen supply would disturb for ATPases to pump Ca^{2+} out of the cytosol. The reduced function of active transporters such as sarcolemma Ca^{2+} ATPase (PMCA) and sarcoplasmic reticulum Ca^{2+} ATPase (SERCA) may contribute to an increase in the cytosolic Ca^{2+}.

4.11 Conclusions

A ligand is usually a small molecule that produces a signal by binding specifically to a site called a receptor on a target protein. Ligands may be hormones, neurotransmitters, second messengers, or substrates of the enzyme. Binding between a ligand and its receptor results from intermolecular forces such as electrostatic forces or van der Waals forces, which cause the conformational change of the target protein. The binding typically changes the three-dimensional shape of the target protein, which switches on various physiological effects depending on the target protein. The target protein may be enzymes, ligand-gated ion channels, or G protein-coupled transmembrane proteins. The binding occurs over an infinitesimal range of time. It is reversible, and the dissociation is fast.

Redox reactions are characterized by the shift of electrons from the reducing agent at the lower electric potential to the oxidizing agent at the higher electric potential. Oxygen atom attracts electrons strongly toward itself, and two electrons from NADH tends to be pulled by the oxidoreductases in the complex I, III, and oxidase in the complex IV, releasing protons into the intermembrane space through the complexes, but not be pulled toward NAD^+, and the negatively charged oxygen atom after accepting two electrons combines with two protons from the matrix, to produce water by cytochrome *c* oxidase.

The electrochemical potential of the protons produced through a series of electron transport in mitochondria drives two molecular motors of F_0F_1 ATP synthase. Without sufficient oxygen, the electron transfer stops, and ATP synthesis comes to a halt when the electrochemical potential of the proton diminishes, and then the enzyme works in the reverse direction as a proton pump using a high-energy phosphate detached from ATP hydrolysis. The resultant ATP deficiency may disturb for Na^+-K^+ ATPases to pump Na^+ out of the cytosol.

References

[1] R.G. Pearson, Absolute electronegativity and absolute hardness of Lewis acids and bases, J. Am. Chem. Soc. 107 (24) (1985) 6801–6806, https://doi.org/10.1021/ja00310a009.

[2] L.C. Allen, Electronegativity is the average one-electron energy of the valence-shell electrons in ground-state free atoms, J. Am. Chem. Soc. 111 (25) (1989) 9003–9014, https://doi.org/10.1021/ja00207a003.

[3] D. Okano, R. Iino, H. Noji, Rotation and structure of F_0F_1-ATP synthase, J. Biochem. 149 (6) (2011) 655–664.

[4] P. Atkins, L. Jones, Chemical Principles, fourth ed., Freeman, New York, 2008 (Chapter 7).

[5] D. Voet, J.G. Voet, Biochemistry, John Wiley & Sons, Hoboken, NJ, 2004 (Chapter 22).

[6] D.C. Crossman, The pathophysiology of myocardial ischaemia, Heart 90 (5) (2004) 576–580.

[7] T. Kalogeris, C.P. Baines, M. Krenz, R.J. Korthuis, Cell biology of ischemia/reperfusion injury, Int. Rev. Cell Mol. Biol. 298 (2010) 229–317, https://doi.org/10.1016/B978-0-12-394309-5.00006-7.

[8] M. Vassalle, C. Lin, Calcium overload and cardiac function, J. Biomed. Sci. 11 (2004) 542–565.

[9] S. Sanada, I. Komuro, M. Kitakaze, Pathophysiology of myocardial reperfusion injury: preconditioning, postconditioning and translational aspects of protective measures, Am. J. Phys. 301 (2011) H1723–H1741.

CHAPTER 5

Resting electric potential and ionic currents in the cell

5.1 Introduction

Along with the continuous actions of Na^+-K^+ ATPase, the membrane that is impermeable to Na^+ and permeable to K^+ at a resting state causes the equilibrium electric potential in the cell, which is called the resting membrane potential or resting voltage. The ATP-driven Na^+-K^+ pump continuously drives three Na^+ ions out of the membrane against the concentration gradient while at the same time pumping two K^+ in from the outside. At a resting state, most of the Na^+ cannot pass through the membrane due to the few Na^+ leak channels while most of the K^+ can pass through. As a result, the concentration of Na^+ is higher outside the cell, whereas the concentration of K^+ is higher inside the cell. This process is continuous and requires high energy from ATP hydrolysis. The Na^+-K^+ ATPase, along with the effects of limited diffusion of Na^+ through the resting membrane, plays a major role in maintaining the resting potential in the cell membrane.

In addition to the resting potential, the small amplitude of rhythmic slow wave activity is seen in both excitable and nonexcitable cells, which may suggest Ca^{2+} oscillation in the cytosol. The common pattern of Ca^{2+} signaling seen in the cells is thought to be the periodic releases of the stored Ca^{2+} from the endoplasmic reticulum (ER) [1] or the SR. The cells can synchronize by spreading waves, producing coordinate events such as the rhythmic beating of cilia in airway epithelial cells [2], cardiomyocytes, and intestinal myocytes [3,4].

When stimulated, excitable cells (neurons, muscles, and some secretory cells in glands) initiate specific electrochemical spikes called the action potential, and most nonexcitable cells can undergo membrane voltage changes in response to environmental or intracellular stimuli.

Modeling Electrochemical Dynamics and Signaling Mechanisms in Excitable Cells with Pathological Case Studies
https://doi.org/10.1016/B978-0-323-98803-2.00005-5

5.2 Diffusion of ions

We see that diffusion begins when solids dissolve in a liquid, different gases or liquids mix, solutes such as sodium ions are injected in the fluid of the axon, or gases are dissolved in the blood. Diffusion is defined as the spread of solute moving from the area of higher concentration to the area of lower concentration along the concentration gradient. The concentrated area can be made by the injection of solutes in a solvent in the small container such as a cell with leak ion channels. Continuous bumping among different kinds of molecules produces a pressure fluctuation which drives them down from the area of higher concentration to the area of lower concentration. A higher concentration gradient will induce a higher probability of a collision between the same kind of molecules that replaces one of the positions with a different kind of a neighbor molecule, which causes mixing and proceeds diffusion. Exchanging of positions between the same kind of molecules occurs more often as concentration gradient goes down over time, thereby reducing the speed of diffusion. The speed of mixing is fast at first and gradually goes down. The speed of diffusion is proportional to concentration gradient of a certain kind of molecule, and it is inversely proportional to the resistivity of the tissue. Diffusion will stop when its concentration or its partial pressure of a certain kind of molecule becomes the same anywhere in the container, where entropy becomes maximized.

The speed of diffusion, called flux or current, is proportional to concentration gradient and temperature. And it is inversely proportional to friction which is a function of the radius of the molecule and the viscosity of fluid if diffusible molecules are nonionic or nonpolar. Gases such as oxygen, carbon dioxide, and nitrogen, and small nonpolar molecules such as steroid hormones can diffuse through the phospholipid bilayer of the cell membrane along their concentration gradients. Because of the high friction, the flux through the phospholipid bilayer is slow. If the solute is an ion or polar molecule, the speed of diffusion depends on the ion conductivity of the tissue and charges of the solute. Ions and polar molecules cannot pass through the phospholipid bilayer and the scar tissue that have extremely low ion conductivity. And to be able to pass through the membrane, they need channels or transporters which are made of columnar proteins firmly inserted into the phospholipid bilayer. Diffusion through channels or transporters is called facilitated diffusion. Because of the low friction, the flux of facilitated diffusion is faster than that of simple diffusion.

A. Fick introduced flux as a measure of the rate of diffusion of spherical nonionic or nonpolar molecules. Physically, flux, i.e., current, is interpreted as the amount of matter passing through per unit time (s).

In 1855, *the first law of diffusion* for one direction was published.

$$J\left(\mathrm{mol/m^2s}\right) = -D\frac{d[\mathrm{C}]}{dx} = -\left(\frac{kT}{f}\right)\frac{d[\mathrm{C}]}{dx} \tag{5.1}$$

where flux per unit area (m^2) is known as the current density and denoted by J (mol/m^2s). The concentration of the solute, [C] depends on the position (x) and D is called the diffusion constant that consists of the absolute temperature (T), and the friction coefficient (f), and the Boltzmann constant (k). The minus sign in Eq. (5.1) indicates that the molecules move from a higher concentration to a lower concentration.

(1) For spherical nonionic or nonpolar molecules like oxygen and nitrogen, Stokes' law is

$$f = 6\pi r\rho$$

where r is the radius of the molecule, ρ is the viscosity of the fluid.

(2) For spherical ionic or polar molecules, the friction coefficient should be replaced with the following equation.

$$J_{\mathrm{con},x}(\mathrm{A/m^2}) = -\left(\frac{neN_\mathrm{A}kT}{f}\right)\frac{d[\mathrm{C}]}{dx}$$

$$f = n^2e^2N_\mathrm{A}[\mathrm{C}]/\sigma$$

where e denotes the charge on an electron, n is the ion valence (the number of charges carried by an ion, e.g., +1 for K^+, +2 for Ca^{2+}, and −1 for Cl^-), N_A is Avogadro's number, [C] is the concentration of the diffusing molecules, and σ is the ion conductivity of the tissue. The ampere is abbreviated as A (1 A = 1 C/s) (see Section 5.7).

The diffusion constant of the living cell can be written as

$$D = \frac{\sigma kT}{ne[\mathrm{C}]} \tag{5.2}$$

Ions cannot pass through the scar tissue or lipid layer in the cell membrane, which cannot conduct any ionic current at all, where σ is almost zero. If solvated solutes pass in a narrow passage of leak channels, an interaction between the solvated ion and the negatively charged interior protein wall of channels may occur. Some ion channels are opened through the conformational change of the channel protein that occurs by binding of the ligand to its receptor, by stimulating voltage sensors, or mechanosensitive sensors. The speed of ion diffusion may depend on the diameter of the passage and the diameter of the solvated ion (see Section 5.4). Fick's law of diffusion can explain the mixing and dilution of solutes dynamically and calculate the speed of the spread of ions.

5.3 Leak channels

An ion channel that always opens even in the cell membrane under rest is called a leak channel; however, the number of leak channels for each kind of ion is different. These channels are composed of a bundle of transmembrane α helices, which are found in the plasma membrane of almost all animal cells. Each different kind of ion leak channels selects and passes only one kind of ion. The neuronal membrane is super permeable to potassium ions, and potassium ions leak out of the neuron through many potassium leak channels. K^+ leak channels are made up of the four subunits of columnar proteins. Each single channel consists of four transmembrane domains (Fig. 5.1) and a filter with one pair of the sequence of five electronegative oxygens in carbonyl groups (Fig. 5.3). Potassium leak pathways are active at membrane potential below the firing threshold and play an important role in expediting repolarization during the last phase of an exciting state [5].

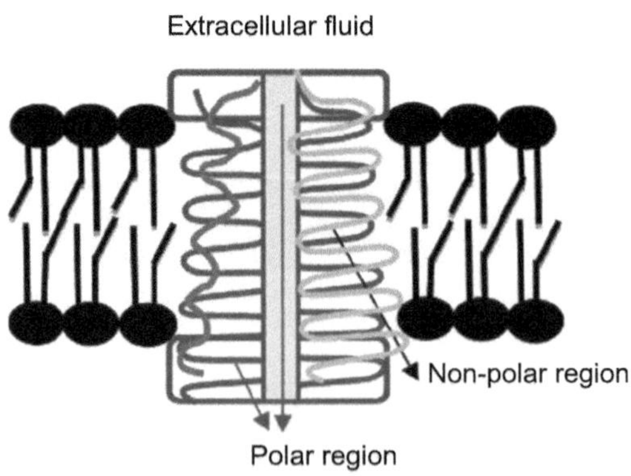

Fig. 5.1 K^+ leak channels are made up of the four subunits of columnar proteins. Each single channel consists of four transmembrane domains and a filter with one pair of the sequence of five electronegative oxygens in carbonyl groups. The hydrophilic regions which interact with water align on both sides of the columnar channel protein and the inner part of the membrane, while the hydrophobic regions located inside the membrane interact with phospholipid tails and cholesterols.

However, the repolarizing neuronal membrane is hardly permeable to sodium ions, that is, most of the sodium ions discharged into to the cell in the exciting state cannot leak out of neurons due to the few numbers of sodium leak channels and higher sodium ion concentration outside of the cell. Na^+-K^+ ATPase takes over pumping sodium ions out against the electrochemical gradient. The shape of the channel protein in the membrane is columnar. If we examine its sequence of amino acids, each hydrophilic and hydrophobic group of amino acids will appear to be at the same distance apart alternately. The hydrophilic regions which interact with water align on both sides of the columnar channel protein and the inner part of the channel pore, while the hydrophobic regions located inside the membrane interact with phospholipid tails and cholesterols (Fig. 5.1).

5.4 Ion selectivity of leak channels

Salts such as NaCl and KCl are dissolved in water because of the bonds separated by the electrostatic force between the elements of the salts and polar water molecules. Na^+ and K^+ become surrounded by polar water molecules in an aqueous solution. Such ions are said to be solvated. Partially negatively charged oxygen atoms of water will surround sodium or potassium ions, which form solvation shell around sodium or potassium ions and make them electrically neutral. The attractive force between the oppositely charged ions is attenuated by the water dipoles. That is why the oppositely charged ions tend to be kept separated and solved in water. Electrostatic force is needed to separate the hydration shell from the ions.

The negatively charged oxygens of carbonyl groups and threonine side chain of membrane proteins of the cell will disturb the hydration shell around the positively charged metal ions. They remove some water molecules around them, making the total size smaller. On the other hand, the negative membrane proteins will repel the hydration shell of negatively charged molecules like nucleic acids. The solvation shell around K^+ in water is shown in Fig. 5.2.

Without ion channels, ions cannot pass through the cell membrane. The sodium ion channel is made of the columnar protein with a tiny pore that is just wide enough to accommodate a sodium ion and one associated water molecule to pass through. The difference of ionic radius between the sodium and potassium ions seems to be insignificant. The atomic radius of K^+ is 138×10^{-12} m while that of Na^+ is a little bit smaller, 102×10^{-12} m. However, the pore of sodium ion channels is too small for the solvation shell

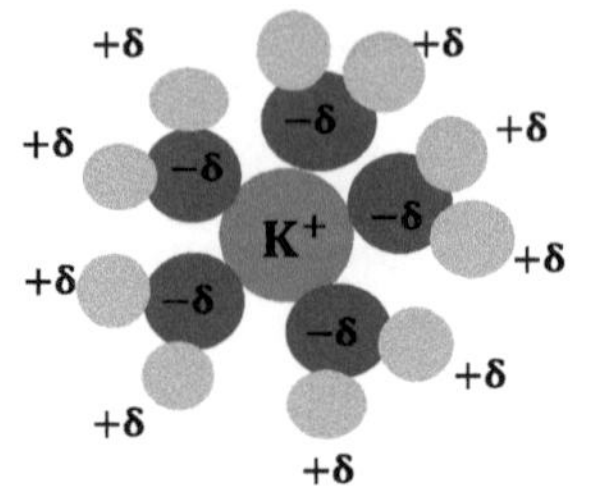

Fig. 5.2 Solvation shell around K^+ in water. The atomic radius of K^+ is 138×10^{-12} m, that of O is 73×10^{-12} m, and that of H is 52.9×10^{-12} m. K^+ ions become surrounded by polar water molecules in aqueous solution. Such ions are said to be solvated. The partially negatively charged oxygen atoms of water will surround the potassium ions, which make solvation shell around potassium ions and make them electrically neutral. The attractive force between the oppositely charged ions is attenuated by the electric field produced by the water dipoles.

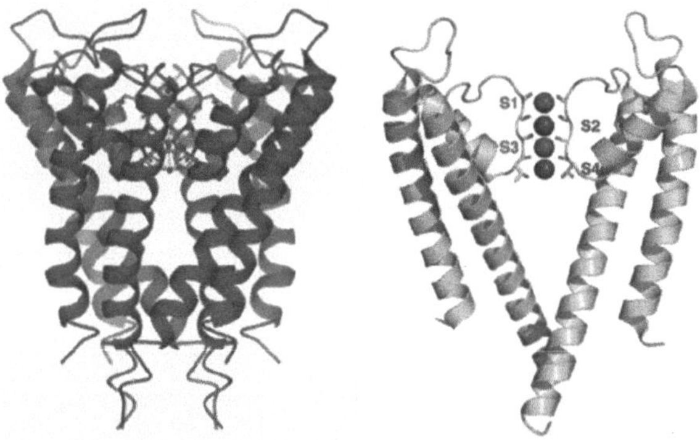

Fig. 5.3 The tetramer of K^+ channel proteins is displayed. The selectivity filter is formed by one pair of the sequence of five electronegative oxygens in carbonyl groups. Potassium ions (occupying the S2 and S4 sites) and oxygen atoms of water molecules (S1 and S3) are shown. Potassium ions with one or two water molecules can just pass through the channel. The atomic radius of K^+ is 138×10^{-12} m, whereas that of Na^+ is 102×10^{-12} m. Smaller Na^+ may have the hydration shell of more than three water molecules and cannot go through.

around the potassium ions. Potassium channels can pass K^+ in the presence of Na^+. Why cannot the smaller Na^+ go through the potassium channel? Potassium channels have the filter formed by one pair of the sequence of five electronegative oxygens of carbonyl groups (>C=O) (Fig. 5.3) [6], selectively fitting for the larger size of potassium ions. It is cation-attractive and allows passage of only potassium ion with one or two associated water molecules. The potassium channel with the filter can accommodate potassium ions with one or two water molecules through interactions between the electronegative filter and the hydration shell of the potassium ions, but the filter is thought to be spaced too far apart to disturb the hydration shell of the smaller sodium ion, keeping its hydration shell of more than three water molecules.

Potassium ions (occupying the S2 and S4 sites) and oxygen atoms of water molecules (at S1 and S3) are shown in Fig. 5.3 [7]. Potassium ions with one or two water molecules can just pass through the channel. Smaller Na^+ may have a hydration shell of more than three water molecules and cannot pass through.

5.5 Na^+-K^+ pump driven by ATP

Along with the continuous actions of Na^+-K^+ ATPase, the membrane that is impermeable to Na^+ and permeable to K^+ at a resting state causes the resting membrane potential in the cell. K^+ is 25 times more permeable than Na^+ at a resting state. The Na^+-K^+ pump helps restore the resting membrane potential after an exciting state by the transport process in which sodium ions are pumped out across the cell membrane against the electrochemical gradient using energy supplied by ATP hydrolysis. To maintain the resting membrane potential, cells must keep a low concentration of sodium ions within the cell. The Na^+-K^+ pump is the enzyme called Na^+-K^+ ATPase that pumps $3Na^+$ out and $2K^+$ into the cell against the concentration gradient through the presence of ATP. ATP is attracted to the activation site of Na^+-K^+ ATPase when intracellular three Na^+ ions bind to the enzyme at state 1 of the transport process as shown in Fig. 5.4. The phosphorylation was confirmed experimentally as a negatively charged Asp (Aspartic acid) residue of the enzyme became a highly reactive aspartyl phosphate intermediate [8]. The binding of high-energy phosphate from the hydrolysis of ATP to the activation site causes the conformational change of the enzyme. At state 2, the decreased Na^+ affinity transfer three Na^+ ions to the outside, and the increased K^+ affinity takes two extracellular K^+. The K^+ binding triggers a release of the phosphate from the pump, and then the enzyme returns to the original low energy state 1 as shown in Fig. 5.4. The enzyme is made to act as the pump through alternate changes of the open side of the transporter induced by phosphorylation and dephosphorylation of the activation site. The concentration of Na^+ becomes lower inside the cell as the Na^+-K^+ pump transfers three Na^+ out against the gradient while at the same time two K^+ are pumped in from outside. This process is continuous and

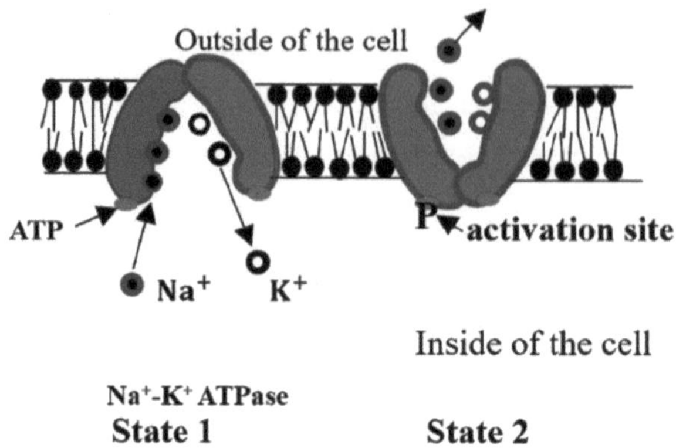

Fig. 5.4 ATP is attracted to the activation site of Na^+-K^+ ATPase when intracellular three Na^+ ions bind to the enzyme at state 1 of the transport process. The binding of high-energy phosphate from the hydrolysis of ATP to the activation site causes the conformational change of the enzyme. The enzyme is made to act as the pump through alternate changes of the open side of the transporter induced by phosphorylation and dephosphorylation of the activation site. At state 2, the decreased Na^+ affinity transfer three Na^+ ions to the outside, and the increased K^+ affinity takes two extracellular K^+.

requires ATP hydrolysis for the successive release of high energy phosphate. The ATP-driven Na^+-K^+ pump works against the electrochemical gradients.

As mentioned, the binding of Na^+ proceeds to attract ATP to the activation site of the enzyme. The formation of aspartyl phosphate intermediate of the enzyme causes the conformational change of the enzyme while a decrease in Na^+ affinity with the binding site of the enzyme in turn evokes the expelling of three Na^+ ions to the outside. Now, extracellular two K^+ ions can bind to the pump protein which has gained an increase in K^+ affinity. The K^+ binding triggers hydrolysis of the phosphate residue and releases the phosphate from the pump. Then the enzyme returns to its original shape and restores an increase in Na^+ but a decrease in K^+ affinity with the binding sites. Thus, the two K^+ ions can leave the pump and enter the cell. The Na^+ binding sites are now ready to bind intracellular three Na^+ ions again.

At state 1, the shape of the enzyme has an inward-facing side with high affinity for Na^+. The enzyme is not phosphorylated at state 1.

At state 2, the shape of the enzyme has an outward-facing side with high affinity for K^+. The enzyme is phosphorylated at the activated state 2.

The permeability of the membrane to Na^+ at a resting state is extremely low compared to that of K^+. Most of the Na^+ pumped out cannot return. However, in excitable tissues, when the potential is raised above the threshold, −55 mV, the voltage-gated Na^+ channels open and allow Na^+ to burst in. Without sufficient oxygen supply, the deficiency of ATP may cause an intracellular Na^+ increase due to the dysfunction of ATPase.

5.6 Resting membrane potential caused by Na^+-K^+ ATPase and selective ion permeability

Different kinds of ions have their own membrane permeability due to the own number of leak channels for each kind of ion. The ratio of the number of leak channels for K^+, Na^+, and Cl^- is 1.00:0.04:0.45 at a resting state. If the same concentrations of K^+ and Na^+ are separated by the membrane as shown in Fig. 5.5, the number of K^+ that can pass through the membrane is 25 times more than the number of Na^+ that can pass at the same time. At rest, sodium ions hardly pass through the membrane due to the few sodium leak channels, and then considerable amounts of K^+ efflux create the electric potential difference (ΔV) between the inside and outside of the membrane, which pushes some K^+ ions back and finally creates the equilibrium membrane potential. In equilibrium, the speed of the number of permeable ions diffusing out through the membrane becomes equal to the speed of the number of ions being pushed back by the electric potential difference created.

After equilibrium is reached, we can calculate the potential difference between the different sides of the membrane. According to statistical thermodynamics, the number of states per identical solute has been known to be inversely proportional to the concentration (see Section 2.6.2). Gibbs

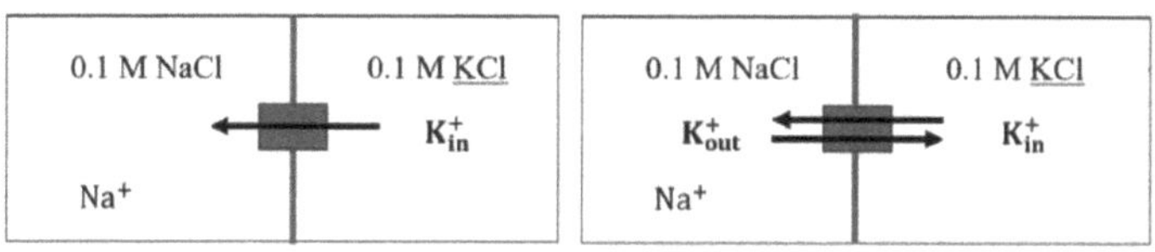

Fig. 5.5 Sodium ions can hardly pass through the membrane due to the few sodium leak channels. However, potassium ions can diffuse through the membrane down along the concentration gradient and then generate the electrical potential difference (membrane potential) that pushes some potassium ions back to the original side where they come from.

free energy per mole due to the concentration difference is described as Eq. (5.3).

$$\Delta G_{\text{dil}} = \Delta H_{\text{dil}} - T\Delta S_{\text{dil}} = -T\Delta S_{\text{dil}} = -RT\ \ln\frac{[\text{higher concentration}]}{[\text{lower concentration}]}$$

$$\Delta G_{\text{dil}} = -RT\ \ln\frac{[\text{K}^+_{\text{in}}]}{[\text{K}^+_{\text{out}}]} \tag{5.3}$$

The total charge of n moles of electrons is $-neN_A$, where N_A is Avogadro's number, $w_{\text{rev, e}}$ denotes the reversible electric work. If $w_{\text{rev, e}}$ is negative, the efflux of K^+ through the membrane generated a power to push some K^+ ions back to the original side from where they come.

$$w_{\text{rev,e}} = -neN_A\Delta V = -nF\Delta V$$

Gibbs free energy per mole of electrons is equal to the reversible electric work (see Section 4.7); ΔV is the difference of electric potential between the inside and outside of the membrane.

$$\Delta V = V_{\text{in}} - V_{\text{out}}$$
$$\Delta G_{\text{elec}} = w_{\text{rev,e}} = -nF\Delta V$$

Here n is equal to the ion valence, that is, the number of charges carried by an ion, e.g., +1 for K^+, +2 for Ca^{2+}, and −1 for Cl^-.

At equilibrium, $\Delta G_{\text{elec}} + \Delta G_{\text{dil}} = 0$

$$-F\Delta V = RT\ \ln\frac{[\text{K}^+]_{\text{in eq}}}{[\text{K}^+]_{\text{out eq}}}$$

The equilibrium potential for potassium (ΔV) is,

$$\Delta V = \frac{RT}{F}\ \ln\frac{[\text{K}^+]_{\text{out eq}}}{[\text{K}^+]_{\text{in eq}}} \tag{5.4}$$

The extracellular fluid contained $[K^+]_{eq} = 4.4$ mM. The intracellular fluid contained $[K^+]_{eq} = 140$ mM. Calculate the equilibrium membrane potential at 37°C under the following conditions if the membrane is 100% permeable to K^+ and 0% permeable to Na^+ and Cl^-.

At equilibrium, the efflux of K^+ by the concentration difference is balanced with the influx by the equilibrium membrane potential.

$$\Delta V = \frac{RT}{nF}\ln K = \frac{(8.315\text{JK}^{-1}\text{mol}^{-1})(310\text{K})}{96{,}490\text{Cmol}^{-1}}\ln\frac{[\text{K}^+]_{\text{out eq}}}{[\text{K}^+]_{\text{in eq}}}$$
$$= 0.0267\ \ln\frac{4.4}{140} = 0.0267 \times (-3.46) = -92\text{mV}$$

5.7 Ionic currents driven by electrochemical forces

Free electrons in a metal wire or mobile ions in an aqueous solution will cause electric currents. In the living organism, the diffusion of ions through ion channels or tissue fibers causes ionic currents. Along with the continuous actions of Na^+-K^+ ATPase, the membrane that is impermeable to Na^+ and permeable to K^+ at a resting state plays a major role in maintaining the resting potential. The ionic current through ion channels and gap junctional channels in the membrane is considered as the sum of current driven by a linear combination of concentration and electric potential gradients of ions as follows.

(1) *Ionic current density caused by concentration gradient:*
Flux is interpreted as the amount of matter passing through per unit time (s).

Flux per unit area (m^2) is known as the current density and is denoted by J (mol/m^{2}s).

According to Fick's first law,

$$\vec{J}_{\text{con}}(\text{mol/m}^2\text{s}) = -D\nabla[\text{C}]$$

$$J_{\text{con},x}\left(\frac{\text{mol}}{\text{m}^3}\times\frac{1\,\text{m}}{\text{s}}\right) = -\left(\frac{kT}{f}\right)\frac{d[\text{C}]}{dx}$$

Since the total charge carried through the unit area ($1\,\text{m}^2$) by a unit length move (1 m) of the ions is $neN_A\frac{d[C]}{dx}$ (C), the electric current generated by $neN_A\frac{d[C]}{dx}$ (C) through the unit area per *second* (s) is

$$J_{\text{con},x}\left(\text{A/m}^2\right) = -\left(\frac{neN_AkT}{f}\right)\frac{d[\text{C}]}{dx}$$

The coulomb (C) is the SI unit of charge. Electric current is measured in coulombs per second, i.e., ampere. The ampere is abbreviated as A (1 A = 1 C/s). The diffusion coefficient (D) is $\left(\frac{kT}{f}\right)$, the friction coefficient is f, k is the Boltzmann constant, T is the absolute temperature, the concentration ([C]) is given by mole per unit volume ($1\,\text{m}^3$), e denotes the charge on the electron, n is the ion valence (the number of charges carried by an ion, e.g., +1 for K^+, +2 for Ca^{2+}, and −1 for Cl^-), and N_A is Avogadro's number.

(2) *Current density caused by electric potential gradient:*

$$\vec{J}_E\left(\text{C/m}^2\text{s}\right) = neN_A[\text{C}]\,\vec{v_d} \tag{5.5}$$

$\vec{v_d}$ is known as the drift speed (m/s). The total charge (C) of the ions passing through the unit area ($1\,\text{m}^2$) in a unit amount of time (s) is $neN_A[C]\ v_d$ provided the concentration is given by mole per unit volume ($1\,\text{m}^3$).

The drift speed can be written as $\vec{v_d}=\vec{a}\,t=\frac{ne}{m}t\vec{E}$ since acceleration (a) is $\frac{ne}{m}E$ where E denotes field intensity, m is the mass of the solvated ion, and t is time. m/t can be replaced with the friction coefficient (f) of the solvated ions moving in the channel.

$$\vec{v_d}=\frac{ne}{f}\vec{E}$$

$$\vec{J}_{\mathrm{E}}(\mathrm{A/m^2})=neN_{\mathrm{A}}[\mathrm{C}]\,\vec{v_d}=\frac{n^2e^2N_{\mathrm{A}}}{f}\vec{E}$$

According to the microscopic Ohm's law, the ionic current density is proportional to the electric field intensity, $\vec{E}$ (force per unit charge):

$$\vec{J}_{\mathrm{E}}=\sigma\vec{E}$$

$$\sigma=\frac{n^2e^2N_A[\mathrm{C}]}{f}$$

where σ is the ion conductivity of the tissue, which is proportional to the concentration of the ion [C] and inversely proportional to friction coefficient (f).

The field intensity (E) is expressed as the gradient of electric potential (V).

$$\vec{E}=-\nabla V$$

$$\vec{J}_{\mathrm{E}}(\mathrm{A/m^2})=\sigma\vec{E}=-\sigma\nabla V$$

Then the x component of $\vec{J}_{\mathrm{E}}$ is

$$J_{Ex}\left(\mathrm{A/m^2}\right)=-\frac{n^2e^2N_{\mathrm{A}}[\mathrm{C}]}{f}\frac{dV}{dx}$$

The ionic current density caused by the concentration gradient and the ionic current density caused by the electric potential gradient are in the opposite direction when crossing the ion channel of the membrane.

$$\vec{J}=\vec{J}_{\mathrm{con}}+\vec{J}_{\mathrm{E}}$$

$$J_x=J_{\mathrm{con},x}(\mathrm{A/m^2})+J_{\mathrm{E},x}(\mathrm{A/m^2})$$

The total current density for each ion through the membrane can be expressed by the modified Fick's equation.

$$J_x=-\left(\frac{neN_{\mathrm{A}}kT}{f}\right)\frac{d[\mathrm{C}]}{dx}-\frac{n^2e^2N_{\mathrm{A}}[\mathrm{C}]}{f}\frac{dV}{dx} \tag{5.6}$$

where J_x is the current density for one dimension, n is the ion valence, and $f = \frac{n^2 e^2 N_A [C]}{\sigma}$.

Opening of the channel decreases the friction coefficient (f) and increases the current density (J), followed by a change in the membrane potential. If positive ions enter the cell, the membrane potential will increase, which in turn works to reduce the positive ion influx.

In equilibrium, the ionic current density caused by the concentration gradient and the current density caused by the electric potential gradient are equal and in the opposite direction, and J_x equals zero. Using Eq. (5.6), the difference of electric potential (ΔV) at equilibrium between the outside and inside of the membrane can be calculated. The concentrations of ions outside and inside of the membrane are denoted by $[C]_{out\ eq}$ and $[C]_{in\ eq}$, respectively.

$$J_{con,x}\left(A/m^2\right) + J_{E,x}\left(A/m^2\right) = 0$$

$$\left(\frac{kN_A T}{f}\right)\frac{d[C]}{dx} = -\frac{neN_A[C]}{f}\frac{dV}{dx}$$

$$\int_{out}^{in} \frac{RT}{[C]} d[C] = -neN_A \int_{out}^{in} dV$$

After integration and adjustment of units, we obtain the Nernst equation.

$$\Delta V = -\frac{RT}{nF}\ln\frac{[C]_{in\ eq}}{[C]_{out\ eq}} = \frac{RT}{nF}\ln\frac{[C]_{out\ eq}}{[C]_{in\ eq}} \tag{5.7}$$

ΔV is called the equilibrium membrane potential, R, i.e., kN_A is the gas constant, F is Faraday's constant (charge of 1 mol of unit charge, 9.65×10^4 C/mol), n is the ion valence, and T is the absolute temperature (K). The changes in the membrane potential (ΔV) are measurable. The ionic current density through the channel in the membrane was proved to be driven by a linear combination of the concentration gradient of the ion and the electric potential gradient.

5.8 Derivation of the Goldman equation

To calculate the actual membrane potential, we should modify the Nernst equation by adding some terms relating to the membrane permeability for each kind of ion. Taking account of all kinds of ions that are not equally permeable, the membrane potential is calculated based on the concentration and the membrane permeability for different kinds of ions that diffuse across it. The Goldman equation is the modified Nernst equation. The electric field (E) and ionic current density (J_x) are assumed to be constant in the channel across the membrane at a resting state. Eq. (5.6) becomes

$$J_x = -\left(\frac{nekN_A T}{f}\right)\frac{d[C]}{dx} - \frac{n^2e^2N_A[C]}{f}\frac{\Delta V}{\Delta x} = -D\frac{d[C]}{dx} + \frac{DnF[C]}{RT}\frac{\Delta V}{L} \tag{5.8}$$

where N_A is Avogadro's number, $R = kN_A$ is the gas constant, $F = eN_A$ is Faraday's constant, and $D = \left(\frac{nekN_A T}{f}\right)$. The membrane potential is $\Delta V = V_{in} - V_{out}$, and L is the thickness of the membrane from $x = 0$ (the channel opening at the inner side of the cell membrane) to $x = L$ (the channel opening at the outer side of the cell membrane), $\Delta x = x_{in} - x_{out} = -L$.

$$dx = \frac{d[C]}{-\frac{J_x}{D} + \frac{nF[C]\Delta V}{RTL}} \tag{5.9}$$

After Eq. (5.8) is separated, using the mathematical technique of separation of variables, Eq. (5.9) is integrated across the membrane from $x = 0$ (inner side of the channel) to $x = L$ (outer side of the channel), assuming J_x is constant in equilibrium.

$$\int_L^0 dx = \int_{out}^{in} \frac{d[C]}{-\frac{J_x}{D} + \frac{nF\Delta V[C]}{RTL}}, \quad \mu = \frac{F\Delta V}{RT}$$

$$-L = \frac{RTL}{nF\Delta V}\left[\ln\left(-\frac{J_x}{D} + \frac{nF\Delta V[C]_{in}}{RTL}\right) - \ln\left(-\frac{J_x}{D} + \frac{nF\Delta V[C]_{out}}{RTL}\right)\right] \tag{5.10}$$

$$-L = \frac{L}{n\mu}\left[\ln\left(-\frac{J_x}{D} + \frac{n\mu[C]_{in}}{L}\right) - \ln\left(-\frac{J_x}{D} + \frac{n\mu[C]_{out}}{L}\right)\right] \tag{5.11}$$

$$-n\mu = \ln\frac{-\frac{J_x}{D} + \frac{n\mu[C]_{in}}{L}}{-\frac{J_x}{D} + \frac{n\mu[C]_{out}}{L}} \tag{5.12}$$

$$e^{-n\mu} = \frac{-\frac{J_x}{D} + \frac{n\mu[C]_{in}}{L}}{-\frac{J_x}{D} + \frac{n\mu[C]_{out}}{L}} \tag{5.13}$$

$$e^{-n\mu}\left(-\frac{J_x}{D}+\frac{n\mu[\mathrm{C}]_{\mathrm{out}}}{L}\right)=\left(-\frac{J_x}{D}+\frac{n\mu[\mathrm{C}]_{\mathrm{in}}}{L}\right) \tag{5.14}$$

$$\frac{J_x}{D}e^{-n\mu}-\frac{J_x}{D}=e^{-n\mu}\left(\frac{n\mu[\mathrm{C}]_{\mathrm{out}}}{L}\right)-\frac{n\mu[\mathrm{C}]_{\mathrm{in}}}{L} \tag{5.15}$$

$$J_x=\frac{e^{-n\mu}\left(\frac{n\mu[\mathrm{C}]_{\mathrm{out}}}{L}\right)-\frac{n\mu[\mathrm{C}]_{\mathrm{in}}}{L}}{\frac{e^{-n\mu}}{D}-\frac{1}{D}} \tag{5.16}$$

$$J_x=\frac{[\mathrm{C}]_{\mathrm{out}}-\frac{[\mathrm{C}]_{\mathrm{in}}}{e^{-n\mu}}}{\frac{1}{n\mu Pe^{-n\mu}}(e^{-n\mu}-1)},P=\frac{D}{L},D=\left(\frac{nekN_{\mathrm{A}}T}{f}\right)\propto\sigma T \tag{5.17}$$

$$J_x=n\mu P\frac{[\mathrm{C}]_{\mathrm{out}}-e^{n\mu}[\mathrm{C}]_{\mathrm{in}}}{1-e^{n\mu}} \tag{5.18}$$

We assumed J_x of each permeable ion is not zero but constant at a resting state. However, the total current density (the sum of J_x for each permeable ion) is zero, and the electric membrane potential (ΔV) is constant at a resting state. The ion conductivity of the membrane is denoted as σ.

$$J_{\mathrm{total}}=0,\quad nP=\frac{n^2ekN_{\mathrm{A}}T}{fL}=\frac{ekN_{\mathrm{A}}T}{fL}>0\ \text{for}\ \mathrm{K}^+,\mathrm{Na}^+,\text{and}\ \mathrm{Cl}^-$$

$$\begin{aligned}&P_{\mathrm{K}}\frac{[\mathrm{K}^+]_{\mathrm{out}}-e^{\mu}[\mathrm{K}^+]_{\mathrm{in}}}{1-e^{\mu}}+P_{\mathrm{Na}}\frac{[\mathrm{Na}^+]_{\mathrm{out}}-e^{\mu}[\mathrm{Na}^+]_{\mathrm{in}}}{1-e^{\mu}}\\&+|P_{\mathrm{Cl}}|\frac{[\mathrm{Cl}^-]_{\mathrm{out}}-e^{-\mu}[\mathrm{Cl}^-]_{\mathrm{in}}}{1-e^{-\mu}}\\&=0\end{aligned} \tag{5.19}$$

$$\begin{aligned}&P_{\mathrm{K}}\frac{[\mathrm{K}^+]_{\mathrm{out}}-e^{\mu}[\mathrm{K}^+]_{\mathrm{in}}}{1-e^{\mu}}+P_{\mathrm{Na}}\frac{[\mathrm{Na}^+]_{\mathrm{out}}-e^{\mu}[\mathrm{Na}^+]_{\mathrm{in}}}{1-e^{\mu}}\\&+|P_{\mathrm{Cl}}|\frac{-e^{\mu}[\mathrm{Cl}^-]_{\mathrm{out}}+[\mathrm{Cl}^-]_{\mathrm{in}}}{-e^{\mu}+1}\\&=0\end{aligned} \tag{5.20}$$

$\mu=\frac{F\Delta V}{RT}$ is constant for all the ions, K^+, Na^+, Cl^-, Ca^{2+}, but n (ion valence) and $|P|$ are different among ions. $nP=\frac{n^2ekN_{\mathrm{A}}T}{fL}=\frac{ekN_{\mathrm{A}}T}{fL}>0$ since $|n|=1$ for K^+, Na^+, and Cl^-. The value of $|P|$ depends on the ion conductivity (σ) at a constant temperature, and it measures the permeability of each kind of ion through the membrane. $P_{Cl} < 0$. We exclude C^{2+} because $n = +2$. The solution to satisfy Eq. (5.19) are

$$P_K[K^+]_{out} - e^{\mu} P_K[K^+]_{in} + P_{Na}[Na^+]_{out} - e^{\mu} P_{Na}[Na^+]_{in} - e^{\mu}|P_{Cl}|[Cl^-]_{out} + |P_{Cl}|[Cl^-]_{in} = 0 \quad (5.21)$$

$$e^{\mu} P_K[K^+]_{in} + e^{\mu} P_{Na}[Na^+]_{in} + e^{\mu}|P_{Cl}|[Cl^-]_{out} = P_K[K^+]_{out} + P_{Na}[Na^+]_{out} + |P_{Cl}|[Cl^-]_{in} \quad (5.22)$$

$$e^{\mu}\left(P_K[K^+]_{in} + P_{Na}[Na^+]_{in} + |P_{Cl}|[Cl^-]_{out}\right) = P_K[K^+]_{out} + P_{Na}[Na^+]_{out} + |P_{Cl}|[Cl^-]_{in} \quad (5.23)$$

$$\frac{F\Delta V}{RT} = \ln \frac{P_K[K^+]_{out} + P_{Na}[Na^+]_{out} + |P_{Cl}|[Cl^-]_{in}}{P_K[K^+]_{in} + P_{Na}[Na^+]_{in} + |P_{Cl}|[Cl^-]_{out}} \quad (5.24)$$

$$\Delta V = \frac{RT}{nF} \ln \frac{P_K[K^+]_{out\ eq} + P_{Na}[Na^+]_{out\ eq} + |P_{Cl}|[Cl^-]_{in\ eq}}{P_K[K^+]_{in\ eq} + P_{Na}[Na^+]_{in\ eq} + |P_{Cl}|[Cl^-]_{out\ eq}} \quad (5.25)$$

where $\mu = \frac{F\Delta V}{RT}$ and potassium, sodium, and chloride permeability are P_K, P_{Na}, and $|P_{Cl}|$, respectively.

Since the ratio of the number of leak channels for K^+, Na^+, and Cl^- is 1.00:0.04:0.45,

$$\Delta V = \frac{RT}{nF} \ln \frac{[K^+]_{out\ eq} + 0.04[Na^+]_{out\ eq} + 0.45[Cl^-]_{in\ eq}}{[K^+]_{in\ eq} + 0.04[Na^+]_{in\ eq} + 0.45[Cl^-]_{out\ eq}}$$

Because Cl^- has a negative charge, $[Cl^-]_{in}$ is in the numerator and $[Cl^-]_{out}$ is in the denominator.

In humans, the extracellular fluid contains Na^+ 140 mM, K^+ 4.4 mM, and Cl^- 105 mM, and intracellular fluid contains Na^+ 15 mM, K^+ 140 mM, and Cl^- 7.0 mM at rest. $P_K : P_{Na} : |P_{Cl}| = 1.00 : 0.04 : 0.45$. Let us calculate the resting membrane potential using the Goldman equation. $T = 37°C$

$$\Delta V = \frac{RT}{nF} \ln \frac{[K^+]_{outeq} + P_{Na}/P_K[Na^+]_{outeq} + P_{Cl}/P_K[Cl^-]_{ineq}}{[K^+]_{ineq} + P_{Na}/P_K[Na^+]_{ineq} + P_{Cl}/P_K[Cl^-]_{outeq}}$$

$$= 0.0267 \times \ln \frac{4.4 + 0.04 \times 140 + 0.45 \times 7.0}{140 + 0.04 \times 15 + 0.45 \times 105}$$

$$= 0.0267 \times \ln \frac{13.15}{187.9} = 0.0267 \times (-2.66) = -71.0\,\text{mV}$$

The calculated result matches the actual resting membrane potential of −70 mV.

5.9 Ca^{2+} oscillation in the RyR2 channel and rhythmic depolarization in the cytosol

The oscillatory release of stored Ca^{2+} from the sarcoplasmic reticulum (SR) or the endoplasmic reticulum (ER) is seen in both excitable and nonexcitable tissue cells.

The IP_3-gated Ca^{2+} channel is in the membrane of the endoplasmic reticulum (ER). When IP3 binds its receptor, it increases the amplitude and rate of the oscillatory release of Ca^{2+} from intracellular stores in the ER into the cytosol, thereby activating various calcium regulated intracellular signals. The oscillatory release of calcium ions produces the tone of smooth muscle (see Section 8.6).

In the heart, the rhythmic release of stored Ca^{2+} has been proposed as the primary source of the cyclic SA node depolarization. Imaging the SA node cells showed Ca^{2+} sparks arising from the SR [9]. The oscillatory release of stored Ca^{2+} from the SR instead of the hyperpolarization activated cyclic nucleotide-gated (HCN) channels may produce the tone of the muscle and was thought to be the basic source of cardiac automaticity according to the experimental results that the complete loss of HCN channels in the mutant mice caused the failure of some impulse initiations in the SA node pacemaker cells, but the intermittent irregular firing of the SA node pacemaker cells was still preserved [10]. I proposed to call the oscillatory release of stored Ca^{2+} the basic source of automaticity [11].

The Ryanodine receptor 2 (RyR2) calcium channel found primarily in the cardiac muscle is leaky even at rest. The oscillatory release of stored Ca^{2+} in the SR through the RyR2 channels at rest was considered as a diffusion phenomenon caused by the high Ca^{2+} concentration difference between the SR and the cytosol. The oscillation was mathematically justified using a model of the RyR2 channel at a resting state. Ca^{2+} diffuses through the RyR2 channels from the SR toward the cytosol, and the concentration in the channel depends not only on the position but also on the time. The concentration of the same point changes with time. The concentration of the stored Ca^{2+} in the SR (Ca^{2+} reservoir) before entering the channel is maintained at a high level of the constant, C_1. The ryanodine channels (the length $= h$) at rest are assumed to have a high friction coefficient (f). The Ca^{2+} concentration in the cytosol is small, $[C]_{in,\ t=0}$. The Ca^{2+} concentration in the cytosol may increase, but the change is assumed to be small.

The concentration of Ca^{2+} in the channel was justified mathematically by solving Eq. (5.26), called Fick's second law of diffusion, which is transformed from Eq. (5.1) that is called Fick's first law of diffusion (see Section 5.2).

$$\frac{\partial[\mathrm{C}]_{x,t}}{dt} = D\frac{\partial^2[\mathrm{C}]_{x,t}}{\partial x^2}, \quad 0 \leq x \leq h \tag{5.26}$$

To solve Eq. (5.26), we use the method of separation of variables since x and t are independent variables.

$$[\mathrm{C}]_{x,t} = X(x)T(t)$$

By substituting this for $[\mathrm{C}]_{x,\ t}$ in Eq. (5.26),

$$\frac{1}{DT(t)}\frac{dT(t)}{dt} = \frac{1}{X}\frac{d^2X}{dx^2} \tag{5.27}$$

"x" and "t" are independent variables. The left-hand side of Eq. (5.27) is a function of t, while the right-hand side does not depend on t but only on x. This is possible if each side of the equation is a constant, $-k^2$. The constant $-k^2$ must be negative since solutions should be within the bounds of possibility. The following two equations are obtained.

$$\frac{d^2X}{dx^2} = -k^2X(x) \tag{5.28}$$

$$\frac{dT(t)}{dt} = -k^2DT(t) \tag{5.29}$$

The solution of $\frac{d^2Z}{dx^2} = -k^2Z(x), k \neq 0$ is

$$Z = e^{\pm ikx} = \cos(kx) \pm i\sin(kx)$$

The real part of Z is $\cos(kx)$ and the imaginary part is $\pm\sin(kx)$. Both are independent solutions of Eq. (5.28) according to the real solution theorem. Using the superposition principle, we can obtain the general solution of Eq. (5.28) as a linear combination of the two independent solutions.

$$X = \mathrm{A}_1x + \mathrm{A}_2 \quad \text{for } k = 0$$

$$X = \mathrm{B}_1\cos(kx) + \mathrm{B}_2\sin(kx) \quad \text{for } k \neq 0$$

The solution of Eq. (5.29) is

$$\int\frac{dT(t)}{T(t)} = -k^2D\int dt$$

$$T = \text{const} \quad \text{for } k = 0$$

$$T = \mathrm{B}_3e^{-k^2Dt} \quad \text{for } k \neq 0$$

Therefore, the concentration of the solute in the channel is

$$[C]_{x,t} = X(x)T(t) = a_1x + a_2 \qquad \text{for } k = 0 \tag{5.30}$$

$$[C]_{x,t} = (A\cos(kx) + B\sin(kx))e^{-k^2Dt} \quad \text{for } k \neq 0 \tag{5.31}$$

According to the superposition principle for the homogeneous differential equation, a linear combination of independent solutions gives the general solution.

$$[C]_{x,t} = a_1x + a_2 + (A\cos(kx) + B\sin(kx))e^{-k^2Dt} \tag{5.32}$$

The concentration at $x = 0$ is always $[C_1]$, and $[C_1] > [C_{in}]$. The concentration ($x > 0$) changes over time.

The initial condition is $[C]_{x=0,\ t=0} = [C_1]$, $[C]_{x=h,\ t=0} = [C]_{in,\ t=0}$. The conditions must satisfy both Eqs. (5.30) and (5.33).

Applying the initial conditions to the solution, Eq. (5.30),

$$[C]_{x=0,t=0} = [C_1] = a_2$$

$$[C]_{x=h,t=0} = [C]_{in,t=0} = a_1h + a_2$$

$$a_1 = \frac{[C]_{in,t=0} - [C_1]}{h} \quad \text{for } k = 0$$

$$[C]_{x,t} = C_1 + \frac{[C]_{in,t=0} - [C_1]}{h}x + (A\cos(kx) + B\sin(kx))e^{-k^2Dt} \tag{5.33}$$

Applying the initial conditions to the solution, Eq. (5.33),

$$[C]_{x=0,t=0} = [C_1] = [C_1] + A, \quad \text{then } A = 0$$

The concentration of calcium ions in the channel is

$$[C]_{x,t} = [C_1] + \frac{[C]_{in,t=0} - [C_1]}{h}x + (B\sin(kx))e^{-k^2Dt}, \quad 0 \leq x \leq h \tag{5.34}$$

k is a real constant. B is a negative real constant that amplifies Ca^{2+} oscillation in the cytosol.

According to the solution, Eq. (5.34), the level of Ca^{2+} in the channel shows a damped oscillation along the straight line with a negative slope as shown in Fig. 5.6. The low diffusion coefficient (D) representing the high friction is a prerequisite for consistent oscillation. In the heart, the phosphorylation of RyR2 by protein kinase A activated by sympathomimetic

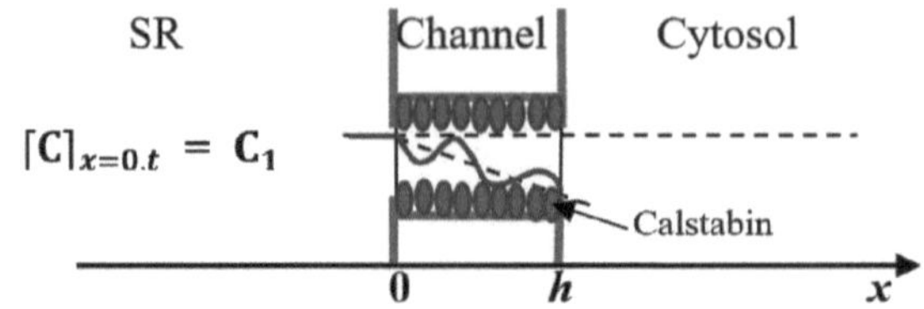

Fig. 5.6 Ca^{2+} oscillation model in the RyR2 channel. The concentration of Ca^{2+} in the channel shows a damped oscillation along the straight line with a negative slope. Phosphorylation of RyR2 and the subsequent dissociation of calstabin may damp Ca^{2+} oscillation due to the increased diffusion coefficient, and Ca^{2+} overload in the cytosol may increase the chance of accidental openings of the channel by allowing calcium ions to bind to RyR2.

stimulation may make the RyR2 channel leakier [12,13], which may damp the Ca^{2+} oscillation due to the increased D in the channel. However, it increases the concentration of calcium ions in the cytosol, which may induce the accidental discharge of massive calcium ions from the RyR2 channel.

The transmembrane protein, sodium-calcium exchanger (Na^+-Ca^{2+} exchanger or NCX), removes Ca^{2+} from the cells. It uses the energy of the electrochemical difference of Na^+ by allowing Na^+ to flow down along its concentration gradient across the plasma membrane in exchange for the countertransport of Ca^{2+}. A single calcium ion is exported for the import of three sodium ions, which results in the depolarization of the membrane potential. The oscillation of Ca^{2+} in the RyR2 channel may generate electrochemical oscillation in the cytosol.

5.10 Conclusions

In the living organism, the diffusion of ions through ion channels or tissue fibers causes electric potential and ionic currents. In equilibrium at rest, the current of each kind of permeable ion is assumed to be not zero but constant while the total sum of current density associated with all permeable ions is zero, and the resting membrane potential is constant. Along with the continuous actions of Na^+-K^+ ATPase, the membrane that is impermeable to Na^+ and permeable to K^+ at a resting state plays a major role in maintaining the resting potential. The leak ion channels have their own selective filter. The difference of the membrane permeability among different kinds of ions is explained by the number of each kind of leak channel. The ratio of the number of leak channels for K^+, Na^+, and Cl^- is 1.00:0.04:0.45 at a resting state. The Goldman equation is derived by taking account of the membrane permeability for each kind of ion. A resting membrane potential is calculated based on the concentration differences of potassium, sodium, and chloride ions across the membrane and the membrane permeability for each kind of ion. Besides resting membrane potential, there is a calcium oscillation in the cell. The release of stored Ca^{2+} from the sarcoplasmic reticulum (SR) or the endoplasmic reticulum (ER) is thought to be the source of oscillation in both excitable and nonexcitable tissue cells.

References

[1] G. Dupont, et al., Calcium oscillations, Cold Spring Harb. Perspect. Biol. (2011), https://doi.org/10.1101/cshperspect.a004226.
[2] K. Barrett, S. Barman, S. Boitano, H. Brooks, Mechanisms of diversity of Ca2+ actions, in: Review of Medical Physiology, twenty fourth ed., McGraw-Hill, New York, 2012, p. 58.
[3] S.D. Koh, K.M. Sanders, S.M. Ward, Spontaneous electrical rhythmicity in cultured interstitial cells of cajal from the murine small intestine, J. Physiol. 513 (1998) 203–213.
[4] L. Thomsen, T.L. Robinson, J.C. Lee, L.A. Farraway, M.J. Hughes, D.W. Andrews, J.D. Huizinga, Interstitial cells of Cajal generate a rhythmic pacemaker current, Nat. Med. 4 (1998) 848–851.
[5] S.A. Goldstein, D.A. Bayliss, D. Kim, F. Lesage, L.D. Plant, S. Rajan, International union of pharmacology. LV. Nomenclature and molecular relationships of two-P potassium channels, Pharmacol. Rev. 57 (4) (2005) 527–540.
[6] S. Varma, S.B. Rempe, Tuning ion coordination architectures to enable selective partitioning, Biophys. J. 93 (4) (2007) 1093–1099.
[7] Y. Zhou, J.H. Morais-Cabral, A. Kaufman, R. MacKinnon, Chemistry of ion coordination and hydration revealed by a K+ channel-Fab complex at 2.0 A resolution, Nature 414 (6859) (2001) 43–48.
[8] D. Voet, J.G. Voet, Biochemistry, third ed., John Wiley & Sons, Hoboken, NJ, 2004 (Chapter 20).
[9] J. Huser, L.A. Blatter, S.L. Lipsius, Intracellular Ca^{2+} release contributes to automaticity in cat atrial pacemaker cells, J. Physiol. 524 (2000) 415–422.
[10] J. Alig, et al., Control of heart rate by cAMP sensitivity of HCN channels, PNAS 106 (29) (2009) 12189–12194, https://doi.org/10.1073/pnas.0810332106.
[11] T. Watanabe, Biophysical Basis of Physiology and Calcium Signaling Mechanism in Cardiac and Smooth Muscle, Academic Press, San Diego, CA, 2018 (Chapter 1).
[12] C. Antzelevitch, A. Burashnikov, Overview of basic mechanisms of cardiac arrhythmia, Card. Electrophysiol. Clin. 3 (1) (2011) 23–45.
[13] A. Duvshani, Structural Studies of the Ryanodine Receptor and Its Binding Protein, Columbia University, Calstabin, 2014, pp. 1–132, https://doi.org/10.7916/D8VD6WM1.

CHAPTER 6

Electrochemical conduction in neurons and neuronal communications

6.1 Introduction

When the cell is at rest, and most of the ion channels are closed except for some leak ion channels, the cell is in a high resistance state called a resting state. When the cell is stimulated, many voltage-gated Na^+ channels may open, and the cell membrane is in a high conductance state called an exciting state. However, negative ion influx or positive ion efflux will lower the membrane potential and cause hyperpolarization which inhibits the cell from generating action potential. Regulation of localized potential in the cell body is done by means of opening Na^+, K^+, Ca^{2+}, and Cl^- channels to control physiological activities.

The opening of ion channels can be attained by bonding the ligand to the receptor on the channel protein, or by means of activation of the voltage or mechanical sensor. On the other hand, the opening of gap junctional channels in myocardium or pacemaker cells is achieved by phosphorylation, Ca^{2+}, and pH [1]. Opening the channels means decreasing friction in the channel. The influx of positive ions raises the membrane potential, and the efflux of positive ions brings down the membrane potential. Ionic current density through ion channels and gap junctional channels in the membrane is considered a result of the sum of current densities driven by a linear combination of concentration and electric potential gradients of ions. When stimulated, excitable cells (neurons, muscles, and some secretory cells in glands) can start a specific electrochemical spike called an action potential, and most nonexcitable cells can also undergo membrane voltage changes in response to environmental or intracellular stimuli. In contrast to efferent neurons, the mechanoreceptor (sensory cell) that is superimposed on the peripheral process of a sensory neuron responds to an external stimulus and cause an action potential [2].

Modeling Electrochemical Dynamics and Signaling Mechanisms in Excitable Cells with Pathological Case Studies
https://doi.org/10.1016/B978-0-323-98803-2.00012-2

The peripheral nervous system collects information from the environment. Afferent neurons are somatic and visceral sensory neurons transmit signals from the sensory receptors in the periphery to the central nervous system (brain and spinal cord). Somatic sensory neurons are responsible for sensory stimulus detection through the specialized mechanoreceptor (sensory cell) that is superimposed on the afferent nerve fiber. Stretch-activated ion channels are located on the sensory cell making the afferent nerve excite. Afferent nerve endings without sensory cells are called free nerve endings. They are less sensitive than the encapsulated afferent fiber and generally function in the perception of pain. Most sensory neurons do not pass directly into the brain, but synapse in the spinal cord allowing prompt reflex actions by activating spinal motor neurons. The brain will receive the sensory input while the reflex is being carried out and the analysis of the signal takes place after the reflex action. The information includes touch, taste, smell, temperature, pain, sound, and vision signals. On the other hand, visceral sensory neurons monitor the levels of carbon dioxide, oxygen, and sugar in the blood; arterial pressure; and the chemical composition of the stomach and gut content.

Efferent neurons lead from the central nervous system back to the periphery to initiate a response to the stimuli. There are two different classes of efferent neurons in the peripheral nervous system; somatic motor neurons and autonomic neurons. Somatic motor neurons contract only skeletal muscle and are responsible for voluntary movements. In the autonomic system, a neuron with a cell body in the central nervous system synapses with a second neuron whose cell body is in a ganglion, i.e., a cluster of neural cell bodies in the periphery. The second neuron extends the axon to the target organ. Within the autonomic nervous system, there are two divisions. In general, the parasympathetic system is active during rest or digestion. The sympathetic system is activated during the fight situations increasing heart rate, releasing glucose from the liver, and dilating pupils.

In sensory neurons, action potentials result from depolarization of the cell body by an external stimulus. In motor neurons, action potentials result from depolarization of the cell body responding to the transmitters released by the presynaptic neuron. However, small injections of depolarizing current in the axon of a neuron can also cause action potential since the axon membrane has voltage-gated channels. Electric stimulations on a somatic motor nerve caused not only a contraction of the innervated muscle fibers but also other skeletal muscle fibers in the same glass container filled with Ringer's solution containing calcium chloride, sodium chloride, and potassium chloride. Adding ACh to the same solution also evokes the muscular contraction like that elicited by stimulation of the motor nerve. The experiments suggested that at the neuromuscular junctions, ACh may be secreted from axon terminals when the action potential reaches the terminals. Some myocytes require no such stimulus to fire. They spontaneously depolarize and fire action potentials at a regular rate. The pacemaker cells in cardiac and intestinal cells provide us with good examples of automaticity.

6.2 Ligand-gated ion channels and neurotransmitters

When binding to neurotransmitters, the ligand-gated ion channels open through the conformational change of the pore protein. These channels control synaptic transmission between two neurons or between a neuron and a muscle fiber. They have the Cys-loop channels, so named because of a large extracellular domain containing Cys loops. Ligand-gated ion channels, also commonly referred to as ionotropic receptors, are a group of transmembrane ion channel proteins. Neurotransmitters are signaling chemicals of the nervous system released from synaptic vesicles formed in the axon terminals of presynaptic neurons, bind specifically to the receptors on the ion channels in the postsynaptic membrane of the target cell such as a skeletal muscle fiber or a postsynaptic nerve cell, and the binding changes the conformation of the pore columnar protein and opens the channel for ions such as Na^+, Cl^-, Ca^{2+}, and K^+ to be able to pass through it. Acetylcholine (ACh) secreted from the presynaptic motor nerve at a neuromuscular junction, ACh secreted from the preganglionic neuron of autonomic nervous in an autonomic ganglion, ACh secreted from postganglionic parasympathetic nerves at a neuromuscular junction and norepinephrine from postganglionic sympathetic nerves, and others such as glutamate and gamma-aminobutyric acid (GABA) in the mammalian brain are examples of neurotransmitters. Inhibitory GABA and excitatory glutamate work together to control the brain's overall level of excitation. The neurotransmitter such as gamma-aminobutyric acid (GABA) in the central nervous system causes hyperpolarization which inhibits nervous excitements in two ways. Ligand-gated Cl^- channel opens when GABA binds to the receptors of the channel in the dendrites of the postsynaptic neuron. The binding of GABA induces the Cl^- influx and causes inhibition of the activity in the central nervous system through hyperpolarization. Binding of GABA to the ligand-gated K^+ channels will cause K^+ efflux, inducing hyperpolarization which inhibits the cell from generating action potential. Glutamate is the most abundant excitatory neurotransmitter, accounting for over 90% of the synaptic connections in the human brain. The α-amino-3-hydroxy-5-methyl-4-isoxazolepropionic acid (AMPA) receptors are on the glutamate-gated ionotropic channels specialized for fast excitation. In many synapses, they produce excitatory postsynaptic potential in their targets in a millisecond after being stimulated. The principal ions gated by AMPARs are sodium and potassium, but sodium influx dominates potassium efflux. *N*-Methyl-D-aspartate (NMDA) receptors are also glutamate-gated ionotropic, but they differ from AMPA receptors in being permeable, when activated, to calcium. Their properties make them particularly important for learning and memory. Alcohol potentiates the sedentary effects of GABA while inhibiting the excitatory aspects of glutamate [3].

6.3 Graded potential in the cell body

Neurons as shown in Fig. 6.1 communicate with one another by the neurotransmitter at the specialized small gap called a synapse. The neurotransmitter released from the terminals of presynaptic neuron diffuses across the synaptic cleft and activates the ligand-gated ion channels on the dendrites or cell body of the postsynaptic neuron to open, causing the localized potentials in the cell. The influx of positive ions such as Na^+, Ca^{2+} raises the membrane potential in the cell and makes the cell excitable. Efflux of positive ions like K^+ and influx of negative ions like Cl^- lowers the membrane potential in the cell, which inhibits the cell from excitement. The postsynaptic potential is the localized potential in the cell body which can be either less negative (−60 mV) or more negative (−80 mV) compared to the resting potential (−70 mV). Once formed, they decay over distance. Graded potentials vary in size which are the results of summation of the excitatory and inhibitory postsynaptic potentials. When the sum of the graded potential reaches the threshold (−55 mV) at the axon hillock, Na^+ influx initiates an action potential via the opening of the voltage-gated sodium channel in the axon.

When the neurotransmitter, acetylcholine (ACh) binds to the receptor on Na^+ channels located mainly in the dendrites of a postsynaptic neuron, the ACh-gated Na^+ channels are opened for sodium ions to move in. As ACh binds, sodium ion channels open via conformational changes of the pore protein, letting more Na^+ in and raising the potential to generate an action potential at the axon hillock. The neurotransmitter such as gamma-aminobutyric acid (GABA) in the central nervous system produces inhibitory postsynaptic potential in their targets through hyperpolarization. The binding of GABA to the ligand-gated Cl^- channel will cause the Cl^- influx. Binding of GABA to the ligand-gated K^+ channels will cause

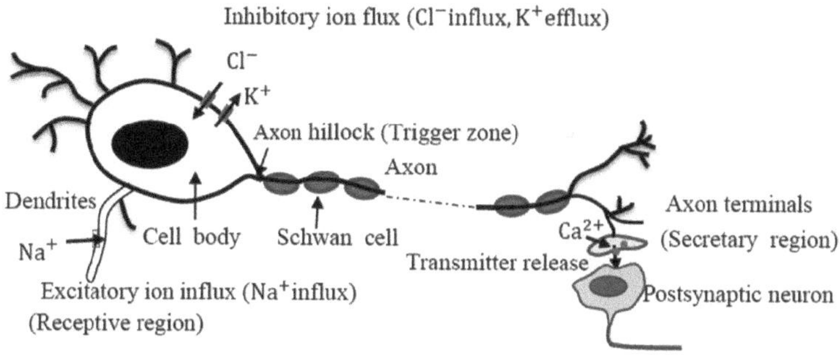

Fig. 6.1 Neurons communicate with one another via neurotransmitters they secrete at the specialized small gap called a synapse. At the axon terminals of the presynaptic neuron, Ca^{2+} influx fires a discharge of neurotransmitters from the terminal when an arriving action potential induces the conformational change of voltage-gated calcium channels to open. Dendrites and a cell body have ligand-gated ion channels which have a receptor for binding to the signaling chemical called a neurotransmitter. Dendrites and a cell body are called receptive regions since they receive stimuli from outside. The neurotransmitters released from the terminals of presynaptic neurons diffuse across the synaptic cleft and activate the ligand-gated ion channels on the dendrites or cell body of the postsynaptic neuron to open, causing the localized potentials called graded potential in the cell. Na^+ influx raises the electric potential in the cell and makes the cell excitable. Cl^- influx and K^+ efflux lower the electric potential in the cell, which inhibits the cell excitement. When the sum of the potential reaches the threshold in the cell body, an action potential is generated at the axon hillock, and transmit the signal unilaterally in the distal direction while any chemical and physical stimuli anywhere on the axon can generate action potentials at the site the stimuli are applied and transmit in both directions. The axon of the peripheral neuron has Schwan cells which increase the speed of action potential transmission by decreasing the electric capacity of the membrane. Axon terminals release neurotransmitters and are said to be a secretory region.

K^+ efflux. In contrast to GABA, glutamate in CNS produces excitatory postsynaptic potential in their targets. The principal ions gated by AMPARs are sodium and potassium, but sodium influx dominates potassium efflux. *N*-Methyl-D-aspartate (NMDA) receptors are also glutamate-gated ionotropic, but they differ from AMPA receptors in being permeable to calcium when activated.

6.4 Switching a neuron between resting state and exciting state

When the cell is at rest and most of the ion channels are closed except for some leak ion channels, the cell is in a high resistance state, and the neuron is said to be in a resting state. Ca^{2+} influx can raise localized potential much faster than Na^+ in the terminals, and at the terminals of the presynaptic neuron, Ca^{2+} influx fires releases of transmitters when an arriving action potential induces the conformational change of voltage-gated Ca^{2+} channels to open. Some ligand-gated ion channels open and produce excitatory postsynaptic potential while some produce inhibitory postsynaptic potential in the cell body. When the graded potential at the axon hillock rises above the threshold, many voltage-gated Na^+ channels open. The cell membrane is said to be in a high conductance state, and the neuron is said to be in an exciting state. The impulse transmits and opens the next Na^+ channel in the axon. Raising the graded potential in the cell body is considered as a switch for starting an action potential. To switch off, we need the neurotransmitter such as GABA in the brain which gives inhibitory postsynaptic potential in the cell body.

Ligands bind to the ligand-gated ion channels and control the excitement of the cell by regulating the electric membrane potential in the cell body. The binding of the neurotransmitters such as ACh and glutamate to the receptors induces excitatory postsynaptic potentials in the cell body and put the switch on to lead the postsynaptic neuron to an exciting state. The neurotransmitter such as GABA in the central nervous system causes inhibitory postsynaptic potential in the cell body and put the switch off to bring the postsynaptic neuron to an inhibitory state and stop the generation of the impulse. In contrast to efferent neurons, the mechanoreceptor (sensory cell) that is superimposed on the peripheral process of a sensory neuron responds to an external stimulus and cause an action potential.

6.5 Structure and function of voltage-gated Na^+ and K^+ channels

Voltage-gated ion channels are transmembrane proteins that are activated by changes in the membrane potential near the channel. The membrane potential alters the conformation of the channel proteins, regulating their opening and closing. Voltage-gated sodium channels have a crucial role in generating an action potential in excitable cells such as neurons and striated muscle cells in response to small increases above threshold in the resting membrane potential.

6.5.1 Voltage-gated Na^+ channels

In excitable cells such as neurons and myocytes, the voltage-gated sodium channel is responsible for the rising phase of action potentials. The channel goes through three different states called resting, active, and inactive states. Even though it would not allow the ions to flow through at both resting and inactive states, the difference between the two states exists with respect to their structural conformation. The pore-forming large α subunit of the voltage-gated channel is composed of four repeated homologous domains (I–IV), each with six transmembrane segments. The S4 transmembrane segment in each homologous domain is considered as the voltage sensor since the electrical field exerts a force on this positively charged amino acid residue, causing the conformational change of the channel. The S5 and S6 transmembrane segments and the short loop between them form the wall of the pore in the center of a symmetrical square array of the four homologous domains. The amino acid residues in the loop are critical for determining the conductance and ion selectivity of the Na^+ channel. When the potential raises above the threshold, suddenly many voltage-gated Na^+ channels open in 1 ms and then the channels close abruptly due to the closure of an inactivation gate. This inactivation gate is thought to be formed by the short intracellular inactivation loop connecting homologous domains III and IV that contains three hydrophobic isoleucine-phenylalanine-methionine residues, which is responsible for the rapid channel closure by entering the intracellular mouth of the pore and binding to an inactivation gate receptor as shown in Fig. 6.2 [4,5].

The local anesthetics reduce not only action potential transmission in neurons by blocking voltage-gated Na^+ channels in the neuronal cell membrane, but also reduce conduction rate in the heart muscle fibers. The local anesthetics have been used not only as local anesthetics, but also as antiarrhythmic drugs.

Amino acid residues in the S6 segment in domains I, III, and IV (Fig. 6.2) are thought to be important for local anesthetic binding. A large hydrophobic amino acid residue such as isoleucine near the extracellular end of this segment creates access of charged local anesthetic drugs from the extracellular fluid to the receptor site for blocking [4]. When the Na^+ channels are blocked by local anesthetics, a greater membrane depolarization is required to bring the Na^+ channels from the resting to the open state. Lidocaine has been used not only as a local anesthetic, but also as an antiarrhythmic drug for the acute intravenous therapy of tachyarrhythmias. Lidocaine has

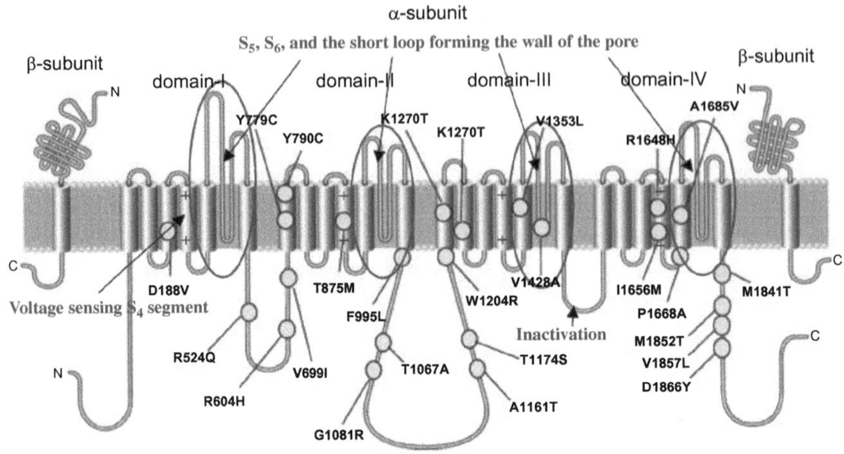

Fig. 6.2 Structure and function of voltage-gated Na^+ channel. The S4 transmembrane segment in each homologous domain is considered as the voltage sensor since the electrical field exerts a force on this positively charged amino acid residue. The short intracellular inactivation loop connecting homologous domains III and IV contains three hydrophobic isoleucine-phenylalanine-methionine residues, which is responsible for the rapid channel closure.

antiarrhythmic effects by blocking voltage-gated sodium channels, thereby increasing threshold and the period of refractoriness along with reducing conduction rate in the cardiac muscle fibers.

6.5.2 Voltage-gated K^+ channels

Voltage-gated potassium channels are transmembrane ion channels and are specific for K^+ and sensitive to voltage changes in the membrane potential. During action potentials, they play an important role in returning the depolarized cell to a resting state. The tetrameric structure of voltage-gated K^+ channels is made of two functionally and structurally independent domains: an ion conduction pore, and voltage-sensor domains. The ion conduction pore is made of four subunits which are arranged symmetrically around the conduction pathway. Voltage-sensor domains are positioned at the periphery of the channel and consist of four transmembrane segments (S1–S4). Structural rearrangement of the voltage-sensor domains in response to changes in the membrane potential, and particularly S4, which includes positively charged amino acids at every third position, results in conformational changes in the conduction pore, which could open or close the ion conduction pathway.

6.5.3 Refractory period caused by inactivation of voltage-gated sodium channels

In neurons, the absolute refractory period is caused by the inactivation of voltage-gated Na^+ channels 1 ms after opening. The closure of the inactivation gate of the Na^+ channel is sustained nearly the entire duration of the action potential. These channels remain inactivated until the membrane hyperpolarizes.

An action potential generated at the axon hillock transmits unilaterally in the distal direction due to the refractory period of the voltage-gated Na^+ channels, in which they cannot be made to open regardless of the membrane potential (see Section 6.5.1). On the other hand, any chemical and physical stimuli anywhere on the axon can generate an action potential at the site the stimuli are applied, and the signal transmits in both directions. Opening of voltage-gated Na^+ channels in the axon by applying the positive voltage above the threshold causes generation of an action potential.

The relative refractory period immediately follows the absolute period. As voltage-gated potassium channels open to terminate the action potential by repolarizing the membrane, the potassium conductance of the membrane increases dramatically. The inactivation of the Na^+ channel is sustained nearly the entire duration of the action potential. As the Na^+ channel channels remain inactivated, efflux of K^+ brings the membrane potential closer to the equilibrium potential for potassium as the ratio of the number of leak channels for K^+, Na^+, and Cl^- approaches to 1:0:0 (see Section 5.6). The efflux of K^+ makes the membrane potential transiently more negative than the resting potential. Until the potassium conductance returns to the resting value, a greater stimulus will be required to initiate the next action potential. The return to the resting potential ends the relative refractory period.

6.6 Action potential in neuron

The membrane potential at a resting state is about −70 mV in humans. It is closer to the equilibrium potential for K^+ since the membrane is 25 times permeable to K^+ than to Na^+. If there are *voltage-gated* Na^+ and K^+ channels in the cell membrane, and the number of open states of voltage-gated channels increases as the potential rises above a certain value, a sudden increase in the permeability for the channels will generate a big potential upstroke called an action potential. The *action potential* arises when the potential is raised above the threshold, −55 mV, and it will reach +30 mV, that is a reversal of the membrane potential with the maximum change of about 100 mV above the resting potential, −70 mV.

If we accept the following assumption, a sharp curve of action potential is easily explainable. When the potential raises above the threshold, suddenly voltage-gated Na^+ channels open and its number of the open state increases sharply in 1 ms until most of them open and then the channels close abruptly, whereas the number of open voltage-gated K^+ channels increases gradually until most of them open in 1.5 ms after the time lag of 0.5 ms and falls gradually as shown in Fig. 6.3. At first, a sharp increase in Na^+ influx raises the potential steeply. And then as the Na^+ influx stops suddenly due to the closure of the inactivation gates in the channels, the time-lagged increase in K^+ outflow restores the potential to the initial state. The sudden closure of the Na^+ channels was examined by analyzing its three-dimensional structure.

Depolarization is defined as the membrane potential at a resting state that becomes less negative toward zero. The change in the membrane potential from −70 to −55 mV is depolarization. By convention, depolarization also includes rising of the membrane potential above zero.

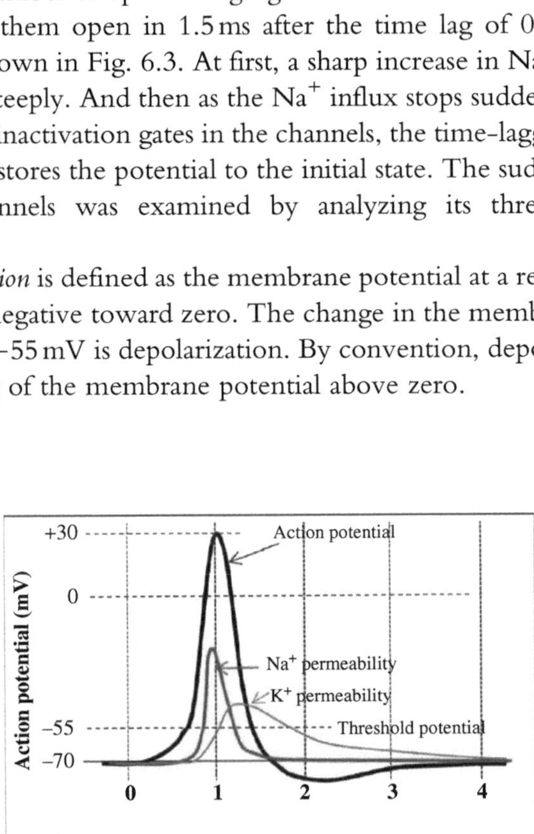

Fig. 6.3 When the membrane potential raises above the threshold, suddenly voltage-gated Na^+ channels open. A sharp spike of an action potential arises along with an abrupt increase in the number of opening of voltage-gated Na^+ channels in the cell membrane. And then the Na^+ influx stops suddenly due to the closure of the channels. The closure of the inactivation gate of the sodium channel is sustained nearly the entire duration of the action potential, while the number of opening of voltage-gated K^+ channels increases gradually and decreases slowly, and K^+ keeps flowing out even after the influx of Na^+ drops sharply. As the Na^+ channels remain inactivated, K^+ moving out of the cell brings the membrane potential to brief hyperpolarization.

Repolarization is defined as returning the membrane potential to the initial resting state.

Hyperpolarization is defined as the membrane potential that becomes more negative than the resting potential. Voltage-gated potassium channels open dramatically to terminate the action potential by repolarizing the membrane. As the Na^+ channels remain inactivated, K^+ moving out of the cell bring the membrane potential closer to the equilibrium potential for potassium as the ratio of the number of leak channels for K^+, Na^+, and Cl^- approaches to 1:0:0 (see Section 5.6). This causes brief hyperpolarization of the membrane, i.e., the membrane potential becomes transiently more negative than the resting potential. Until the potassium conductance returns to the resting value, a greater stimulus will be required to initiate the next action potential.

Why repolarization occurs? As soon as the potential is raised above the threshold, the number of open Na^+ channels rises sharply and then the open channels close abruptly. The closure of the inactivation gate of the sodium channel is sustained nearly the entire duration of the action potential, while the number of open K^+ channels increases gradually and decreases slowly as shown in Fig. 6.3. K^+ keeps flowing out even after the influx of Na^+ drops sharply and the Na^+-K^+ ATPase pump works continuously, which brings back to the initial resting state.

The cell membrane acts like a capacitor (Fig. 6.4). The phospholipid bilayer is a thin insulator separating two electrolytic media, i.e., the extracellular fluid and the cytosol. The membrane capacitance is proportional to the cell surface area (A) and inversely proportional to the width of the membrane including a Schwan cell (d). In the cell, the extracellular and intracellular fluids are the conductors, and the lipid membrane is the

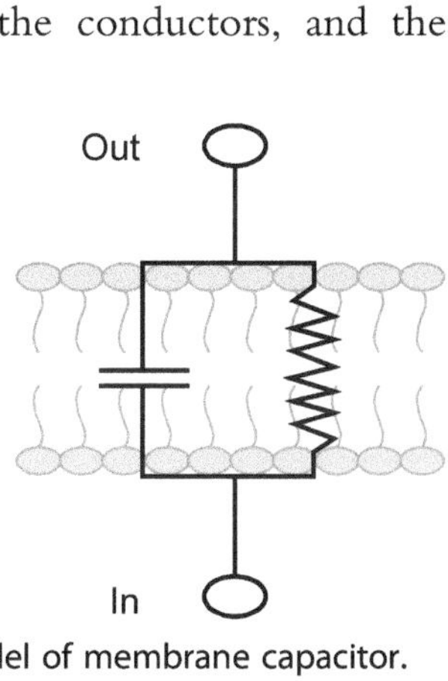

Fig. 6.4 RC circuit as a model of membrane capacitor.

insulator with ion channels. When there is a voltage difference across an insulator, charge (+Q) will build up on the cell surface area (A) outside by Na^+-K^+ ATPase pumping since a little Na^+ current (I) cross the insulator through embedded sodium ion channels with the membrane resistance (R_m) at a resting state. The constant that describes the relationship between the membrane potential (V_m) and the ability of the membrane to store the charge (Q) that builds up is called the membrane capacitance (C_m). Since the cell membrane is made of a nonconducting material, the capacitance of an axon is expressed in terms of the geometry of the membrane as follows.

$$Q = C_m V_m = \left(K\epsilon_0 \frac{A}{d} \right) V_m \tag{6.1}$$

where $\epsilon = K\epsilon_0$ is the permittivity of the dielectric material, and ϵ_0 is called the permittivity of free space. The capacitance is proportional to the dielectric constant, K that is a property of a nonconducting material referred to as dielectrics. Dielectric constant of air is nearly one. The letter A indicates membrane surface area, and d is the width of the membrane including a Schwan cell.

Membrane resistance (R_m) is a measure of how hard ions can move through the membrane. For neurons, the R_m is inversely proportional to the size of the cell with larger A. At rest, a fewer number of Na^+ leak channels and all the closed voltage-gated Na^+ channels make the higher membrane resistance. Membrane potential changes can be calculated using Ohm's law ($V_m = I_m R_m$), where V_m is the membrane voltage in Volts, the I_m is the current through the membrane in Amperes, and the R_m is the resistance in Ohms. The inverse of resistance is conductance (abbreviated as g_m), which is a more convenient way to describe the number of open channels conducting current in a cell. Replacing resistance with conductance, Ohm's law can be written as $V_m = {}^{I_m}/_{g_m}$. When the cell is at rest, and most of the ion channels are closed except leak channels, the cell is described as in a high resistance state. When the membrane is depolarized generating an action potential, many channels are open. Then the cell membrane is said to be in a high conductance state.

$$I_m R_m = -\frac{dq}{dt} R_m = V_m = \frac{q}{C_m} \tag{6.2}$$

When voltage-gated Na^+ channels suddenly open, R_m becomes small. Then sodium ions burst into the axon through the channels causing the positive current which is $-\frac{dq}{dt} > 0$ (the charge build up on the cell surface at rest is

(+ Q) when $t = 0$, and the charge (q) at t decreases by $dq < 0$ per unit time), and the cell will generate an action potential and the positive ions will diffuse in the axon.

$$\frac{dq}{dt} = -\frac{q}{C_m R_m}, \quad \int_Q^q \frac{dq}{q} = -\int_0^t \frac{dt}{C_m R_m} \tag{6.3}$$

$$\ln\frac{q}{Q} = -\frac{t}{C_m R_m} \tag{6.4}$$

$$q = Qe^{\frac{-t}{C_m R_m}} \tag{6.5}$$

where Q is the charge at a resting state, and t is time.

When voltage-gated channel is in a resting state, the channel pore is closed but an inactivation gate is open. The resting membrane potential ($V_{m,\ t=0}$) is created by the pumping of Na^+ by Na^+-K^+ ATPase in addition to the membrane that is impermeable to Na^+ and permeable to K^+.

$$R_m = \infty, \quad I_m = 0, \quad q = Q$$

When the potential becomes above the threshold, suddenly many voltage-gated Na^+ channels open. The voltage-gated channels are in an exciting state, and both the channel pore and gate are open. A large amount of Na^+ charge stored on the surface of the membrane discharges exponentially over time.

$$q = Qe^{\frac{-t}{C_m R_m}}$$

$t > 0$ measures the time after the abrupt opening of the channel.

The channels close abruptly due to the closure of the inactivation gate. Though the channel is still in an exciting state with the pore open, the channel enters the refractory period and remains inactivated nearly entire duration of the action potential. The charge (q) on the surface of the membrane becomes the minimum after the discharge and slowly recovers the charge (Q) as the membrane potential returns to the initial resting state.

$$R_m = \infty, \quad I_m = 0, \quad q \text{ is the minimum at the early stage of the refractory period.}$$

When the width of membrane (d) is smaller, this built-up charge (Q) at the resting V_m becomes larger since the attraction of the opposite charges on the opposite sides of the membrane is large enough to hold larger Q on each side of the cell surface area (A). For neurons, the membrane capacitance is proportional to (1) the size of the cell and (2) inversely proportional to the distance between the conducting materials. Myelinated axons have smaller capacitances than unmyelinated axons of the same size. The capacitance of an axon of a myelinated motor neuron is low due to the width of a Schwan cell around the axon. Eq. (6.5) suggests a larger amount of Na^+ charge stored on the surface of the membrane of the low capacitance (C_m) will discharge in an exciting state with both the channel pore and the gate open, and the amount of discharge will exponentially decrease over time as shown in Fig. 6.5.

Example 6.1 (1) Calculate the membrane capacitance (C_m) of the axon of 1 cm long and the radius of 10 μm. Dielectric constant (K) of the membrane is set to be 3. (2) Estimate change of the concentration of sodium ions in the cell induced by one action potential.

The membrane of the axon is modeled as a cylindrically shaped parallel-plate capacitor. The separation of the plates is the thickness of the membrane which is about 10^{-8} m.

(1) The area of the cylinder of radius (r) and length (l) is
$A = 2\pi r l \approx 6.28 \times (10^{-5}\ \text{m}) \times (0.01\ \text{m}) \approx 6.2 \times 10^{-7}\ \text{m}^2$

$$\text{Capacitance} = K\epsilon_0 \frac{A}{d} \approx 3 \times \left(8.85 \times \frac{10^{-12}\,\text{C}^2}{\text{N}}\ \text{m}^2\right) \frac{6.2 \times 10^{-7}\,\text{m}^2}{10^{-8}\,\text{m}}$$
$$\approx 1.6 \times 10^{-9}\ \text{F}$$

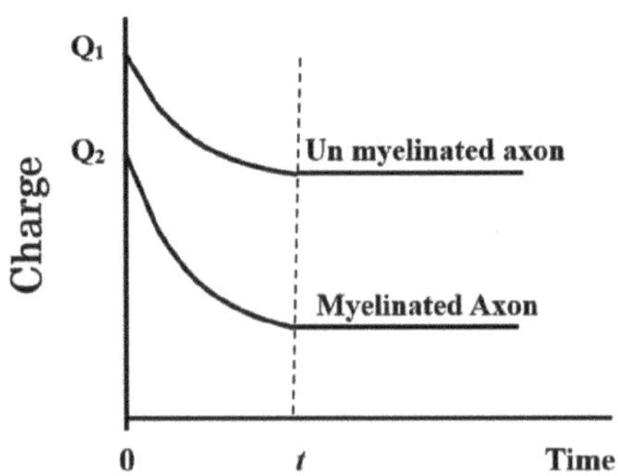

Fig. 6.5 A larger amount of Na^+ charge stored on the surface of the myelinated membrane with the low capacitance (Q_2) will discharge in an exciting state with both the channel pore and the inactivation gate open, and the amount of discharge will exponentially decrease over time. The charge (q) on the surface of the membrane becomes the minimum after the discharge and slowly recovers the charge (Q) as the membrane potential returns to the initial resting state.

(2) Since the voltage changes from −70 to +30 mV, the total change is about 100 mV. The amount of the charge that moves in is

$$\Delta q = C_m \Delta V_m \approx \left(1.6 \times 10^{-9}\,\text{F}\right)(0.1\,\text{V}) = 1.6 \times 10^{-10}\,\text{C}$$

Each sodium ion carries 1.6×10^{-19} C.
The number of Na^+ that flows across the membrane is

$$\Delta q/e \approx 1.6 \times 10^{-10}\,\text{C}/1.6 \times 10^{-19}\,\text{C} = 10^9$$

The volume of the cylindrical axon is

$$\text{Volume} = \pi r^2 l = 3.14 \times \left(10^{-5}\right)^2 \times 0.01 = 3.14 \times 10^{-12}\,\text{m}^3$$

The number of sodium ions contained in 1 m^3 of the intracellular fluid at a resting state is

$$\begin{aligned}[Na^+]_{eq} &= 15\,\text{mM} = 15 \times 10^{-3}\,\text{mol/L} = 15\,\text{mol/m}^3\\ &= 15 \times 6.02 \times 10^{23}\,\text{ions/m}^3 \approx 90 \times 10^{23}\,\text{ions/m}^3\end{aligned}$$

The cylindrical axon contains $90 \times 10^{23} \times 3.14 \times 10^{-12} = 2.83 \times 10^{13}$ sodium ions at a resting state.
About 10^9 sodium ions diffuse in through the voltage-gated sodium ion channels during one action potential passing, which will not change the concentration inside of the axon significantly after mixing. The ions diffuse and propagate down the electrochemical gradient in the axon until it is renewed at the next voltage-gated sodium ion channel.

Example 6.2 Let us calculate the relative permeability of Na^+ to K^+ at the maximum membrane potential of +30 mV. We assume extracellular fluid contains Na^+ 140 mM, K^+ 4.4 mM, and Cl^- 105 mM, and intracellular fluid contains 15 mM of Na^+, 140 mM of K^+, and 7.0 mM of Cl^- at rest, and the concentrations may not change significantly at the exciting state (see Example 6.1).

We assume $P_K : P_{Na} : P_{Cl} = 1.00 : x : 0.45$.

$$\Delta V = 0.030 = \frac{RT}{nF} \ln \frac{[K^+]_{out\,eq} + P_{Na}/P_K\,[Na^+]_{out\,eq} + P_{Cl}/P_K\,[Cl^-]_{in\,eq}}{[K^+]_{in\,eq} + P_{Na}/P_K\,[Na^+]_{in\,eq} + P_{Cl}/P_K\,[Cl^-]_{out\,eq}} \quad (6.6)$$

$$0.030 = 0.0267 \times \ln \frac{4.4 + 140x + 0.45 \times 7.0}{140 + 15x + 0.45 \times 105}$$

$$= 0.0267 \times \ln \frac{7.55 + 140x}{187.25 + 15x}$$

$$1.12 = \ln\frac{7.55 + 140x}{187.25 + 15x}$$

$$e^{1.12} = 3.065 = \frac{7.55 + 140x}{187.25 + 15x}$$

$$574 + 46x = 7.55 + 140x, \quad x = 6.03$$

Relative permeability of Na^+ to K^+ at the maximum potential is 6.03. It was 0.04 at a resting state. At the maximum potential, the relative permeability of Na^+ to K^+ increases 151 times. The action potential is the outcome of changes of the relative permeability in the cell membrane. The increasing number of open Na^+ channels makes the membrane more permeable to sodium ions.

6.7 Velocity of action potential conduction in neurons

Sodium ions enter the cytoplasm through open voltage-gated Na^+ channels in an exciting state, and then the solvated sodium ions start diffusing in the axon. The rate of impulse propagation in neurons depends on the axon diameter and degree of myelination. A larger neuron has smaller axoplasmic resistance (R_i) which makes ions more likely to propagate down the axon at higher velocity. Oligodendrocytes in central nerve fibers and Schwan cells in peripheral nerve fibers make myelin sheath and wrap around segments of the axon. A larger myelinated motor neuron speeds up the propagation of neuronal signals through both smaller axoplasmic resistance (R_i) and a larger amount of Na^+ discharge from the surface of the membrane of the small capacitance (C_m) through many open sodium channels (Eq. 6.5). The signal will wither without another sodium influx in a short distance away.

Nerve fibers are classified according to their diameter and the degree of myelination. Group A fibers are (1) somatic motor fibers that transmit messages about desired movement from the CNS to the skeletal muscles to make them contract, and (2) somatic sensory fibers that carry information like light touch, vibration, pressure, and cutaneous tension to the CNS. They have the largest diameter and thick myelin sheaths, and the speed of impulse conduction goes up to 150 m/s. Group B fibers are lightly myelinated with intermediate diameter. They are (1) visceral sensory fibers that monitor the levels of carbon dioxide, oxygen, and sugar in the blood, arterial pressure and the chemical composition of the stomach and gut content, and

(2) preganglionic autonomic nervous nerve fibers from the CNS. The conduction rate of impulse is about 3–15 m/s. Group C fibers are unmyelinated and have the smallest diameter, average 0.2–1.5 μm in diameter. Group C fibers include postganglionic fibers in the autonomic nervous system (autonomic nervous motor nerves) and nerve fibers at the dorsal roots that carry sensory information from the skin such as pain and temperature fibers. The conduction rate is no more than 2 m/s but is sensitive to the stimulus. The smallest motor neuron with the low membrane capacitance is activated first by an excitatory input while a large motor neuron is recruited later according to the volume of neurons.

Demyelination is the degeneration of the myelin sheath, or the loss of the myelin sheath insulating the axon of neurons. Multiple sclerosis is a demyelinating inflammatory disease of the CNS white matter that displays mononuclear cell infiltration, demyelination, and scarring resulting from the immunological attack. The peripheral nervous system is uninvolved. The cause of multiple sclerosis has not been clearly understood. However, it is known that there is the presence of autoantibodies to myelin oligodendrocyte glycoprotein in the CNS plaque tissue. These antibodies may act with pathogenic T cells to produce the cellular pathology of multiple sclerosis [6]. It is considered an autoimmune disease in which the body's immune system attacks its own tissues. In the case of multiple sclerosis, this immune system malfunction destroys myelin. When the myelin degrades, the conduction of signals along the nerve can be impaired or lost. Symptoms may include numbness or weakness in one or more limbs that typically occurs on one side of your body at a time, and partial or complete loss of vision, usually in one eye at a time, often with pain during eye movement.

Traumatic injury of the myelin sheath will slow the conduction. ATP deficiency may also cause lowering of the conduction since Na^+-K^+ ATPase would not properly work, and the change of the localized potential would be not enough to produce an action potential since the discharge of transmitters from the presynaptic neuron may be decreased.

6.8 Deriving of the cable equation

As voltage-gated Na^+ channels open, the sudden influx of Na^+ causes an action potential. In this model, a current injection is used as a substitution for a sudden influx of Na^+. The cable equation is derived by using the model that a current injected in the axon travels longitudinally through the axoplasm in the axon with small leaky holes. Some of the current will escape

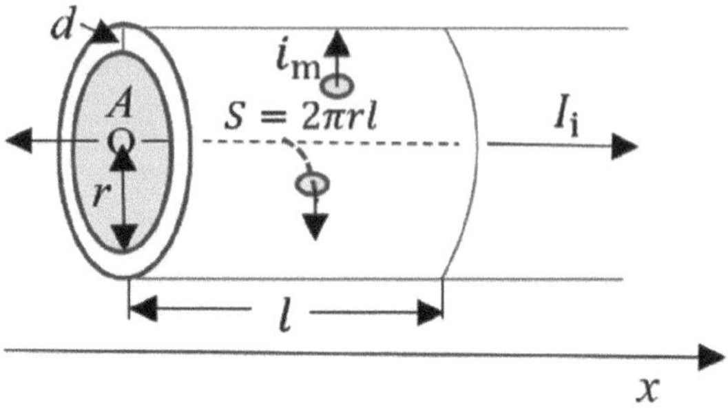

Fig. 6.6 The cable equation is derived by using the model that a current injected in the axon travels longitudinally through the axoplasm in the axon with small leaky holes. Some of the current will escape through membrane ion channels inserted into the membrane and the rest withers over distance and time. The longitudinal axon length that we are considering is denoted by *l*, and *d* is the width of the membrane. *A* is the cross-sectional area of the axon, and O is the origin of internal axoplasmic current. There are ionic channels some currents go through the pores.

through the membrane, and the rest withers shortly over distance and time during transmitting in the axon (Fig. 6.6). Here, resistance (r) and capacitance (c) of the axon of a neuron are measured per unit length (per meter). Thus, r is measured in (Ω/m) and c in farads per meter (F/m). This contrasts with R (Ω) and C (F), which represent the resistance and capacitance, respectively.

$$\mathrm{r} = \frac{\mathrm{R}}{l}, \quad \mathrm{c} = \frac{\mathrm{C}}{l} \tag{6.7}$$

where the membrane capacity per length is $\mathrm{c_m} = \mathrm{C_m}/l = \left(K\epsilon_0 \frac{S}{ld}\right)$, and S is the surface area of the axon, $2\pi r \Delta l\,(\mathrm{m}^2)$, d is the width of the membrane (m), and l is the longitudinal length (m).

The current density (J) is proportional to the electric field (E). Assuming E to be uniform within the length l, $\Delta V_i = El$, where ΔV_i is the change of internal axoplasmic voltage causing the change of the current, ΔI_i.

$$J = \frac{\Delta I_i}{A} = \sigma E = \sigma\left(\frac{\Delta V_i}{l}\right) \tag{6.8}$$

where σ is called the conductivity while ρ is called resistivity. A is the cross-sectional area of the axon, πr^2.

Eq. (6.8) becomes

$$\rho = \frac{1}{\sigma} = \frac{\Delta V_i \cdot A}{\Delta I_i \cdot l}$$

The internal axoplasmic resistance (R_i) is

$$R_i = \frac{\Delta V_i}{\Delta I_i} = \frac{\rho l}{2\pi r^2}$$

A larger neuron has a smaller internal axoplasmic resistance and a larger membrane capacitance. Increasing the size of the neuron lowers the internal axoplasmic resistance more than it increases the membrane capacitance since $R_i \propto r^{-2}$, and $c_m = C_m / l \propto r$.

The internal axoplasmic resistance per unit length, r_i (in Ω/m) is

$$r_i = \frac{\rho}{2\pi r^2} \tag{6.9}$$

The greater the cross-sectional area of the axon, πr^2, the greater the area for the charged ions to pass through its axoplasm, and the lower the axoplasmic resistance.

$$\Delta V_i = -\Delta I_i R_i = -\Delta I_i r_i \Delta x \tag{6.10}$$

where the negative sign means that the current goes down the potential gradient from the point ($x = 0$) where the voltage is applied. The voltage will decrease together with the distance as the current flows down, where $\Delta I_i > 0$.

$$\frac{dV}{dx} = -\Delta I_i r_i$$

$$\frac{1}{r_i}\frac{dV}{dx} = -\Delta I_i \tag{6.11}$$

In the injected axon, electrical potential V_0 creates the current which travels longitudinally through the axoplasm. Some of the current will escape through membrane ion channels inserted into the membrane and wither along with the distance and time. The greater the number of leak ion channels, the faster the ions will escape from the axon. If i_m is the current escaping out from the axon through the membrane per unit length (per meter) of the axon, then the change of the current in the axoplasm, ΔI_i is proportional to the distance, Δx, from the position $x = 0$ can be written as

$$\Delta I_i = -i_m \Delta x \tag{6.12}$$

$$\frac{dI_i}{dx} = -i_m \tag{6.13}$$

i_m can be expressed by the total of the charge going out through the membrane and the flow of stored charge back to the membrane on the side of the cytoplasm from the membrane capacitor. The current (i_c) is usually referred to as displacement current.

The constant that describes the relationship between the membrane potential (V_m) and the ability of the membrane to store the charge (Q) that builds up per unit length of the axon is called the membrane capacitance per length (c_m). $\Delta Q = c_m \Delta V_m$, Q is the charge at a resting state, q is the charge at an active state, and t is time,

$$i_c = \frac{dq}{dt} = c_m \frac{dV}{dt} \tag{6.14}$$

The current that passes the membrane is the sum of the current going out the membrane channel ($i_{outward}$) and the current of the stored charge into the axon through the channel (i_c) as the membrane potential (V) going down,

$$i_m = i_{outward} + i_c \tag{6.15}$$

$$i_{outward} = \frac{V}{r_m}$$

where r_m is membrane resistance inversely proportional to the number of open ion channels.

From Eqs. (6.13) and (6.15),

$$\frac{\partial I_i}{\partial x} = -i_m = -\left(\frac{V}{r_m} + c_m \frac{\partial V}{\partial t}\right) \tag{6.16}$$

Considering both Eqs. (6.11) and (6.16) gives the first version of a cable equation.

$$\frac{1}{r_i}\frac{\partial^2 V}{\partial x^2} = \left(\frac{V}{r_m} + c_m \frac{\partial V}{\partial t}\right) \tag{6.17}$$

By multiplying r_m on both sides of Eq. (6.17),

$$\frac{r_m}{r_i}\frac{\partial^2 V}{\partial x^2} = c_m r_m \frac{\partial V}{\partial t} + V \tag{6.18}$$

6.9 Damped oscillatory voltage change in the exciting axon

To solve Eq. (6.18), the method of separation of variables is considered since x and t are independent variables.

$$V(x,t) = X(x)T(t) \tag{6.19}$$

$$\frac{r_m}{r_i}\frac{\partial^2 X(x)T(t)}{\partial x^2} = \frac{r_m}{r_i}T(t)\frac{\partial^2 X(x)}{\partial x^2} \tag{6.20}$$

$$c_m r_m \frac{\partial X(x)T(t)}{\partial t} + V(x,t) = c_m r_m X(x)\frac{\partial T(t)}{\partial t} + X(x)T(t) \tag{6.21}$$

$$\frac{r_m}{r_i}T(t)\frac{\partial^2 X(x)}{\partial x^2} = c_m r_m X(x)\frac{\partial T(t)}{\partial t} + X(x)T(t) \tag{6.22}$$

Eq. (6.22) is divided by $X(x)T(t)$ and multiplied by $\frac{r_i}{r_m}$,

$$\frac{1}{X(x)}\frac{\partial^2 X(x)}{\partial x^2} = \frac{c_m r_i}{T(t)}\frac{\partial T(t)}{\partial t} + \frac{r_i}{r_m} \tag{6.23}$$

The left-hand side of Eq. (6.23) is a function of x, while the right-hand side does not depend on x but only t. This is possible if each side of the equation is a constant, $-k^2$. The constant $-k^2$ must be negative since solutions should be within the bounds of possibility.

$$\frac{1}{X(x)}\frac{\partial^2 X(x)}{\partial x^2} = \frac{c_m r_i}{T(t)}\frac{\partial T(t)}{\partial t} + \frac{r_i}{r_m} = -k^2 \tag{6.24}$$

The following two equations are obtained.

$$\frac{\partial^2 X(x)}{\partial x^2} = -k^2 X(x) \tag{6.25}$$

$$\frac{1}{T(t)}\frac{\partial T(t)}{\partial t} = -\frac{k^2}{c_m r_i} - \frac{1}{r_m c_m} = -{k_1}^2 \neq 0 \tag{6.26}$$

The solution of $\frac{\partial^2 Z(x)}{\partial x^2} = -k^2 Z(x),\ k \neq 0$ is

$$Z = e^{\pm kix} = \cos(kx) \pm i\sin(kx) \tag{6.27}$$

The real part of Z is $\cos(kx)$ and the imaginary part is $\pm\sin(kx)$. Both are independent solutions of Eq. (6.25) according to the real solution theorem. Using the superposition principle, we can obtain the general real solution of Eq. (6.25) as a linear combination of the two independent solutions.

$$X = A_1 x + A_2 \qquad \text{for } k = 0.$$

$$X = B_1\cos(kx) + B_2\sin(kx) \quad \text{for } k \neq 0 \tag{6.28}$$

The solution of $\frac{1}{T(t)}\frac{\partial T(t)}{\partial t} = -\frac{k^2}{c_m r_i} - \frac{1}{r_m c_m} = -{k_1}^2 \neq 0$ is

$$T = B_3 e^{-{k_1}^2 t} \quad \text{for } k_1 \neq 0 \tag{6.29}$$

Therefore, the electric potential in the axon is

$$V(x,t) = X(x)T(t) = (A_3x + A_4)e^{-k_1^2 t} \quad \text{for } k = 0 \tag{6.30}$$

$$V(x,t) = (b_1\cos(kx) + b_2\sin(kx))e^{-k_1^2 t} \quad \text{for } k \neq 0 \tag{6.31}$$

According to the superposition principle for the homogeneous differential equation, a linear combination of independent solutions gives the general solution.

$$V(x,t) = (A_3x + A_4 + b_1\cos(kx) + b_2\sin(kx))e^{-k_1^2 t} \tag{6.32}$$

The initial condition is $V_{x=0,\ t=0} = V_0$ for Eq. (6.30).

$$V_0 = A_4 \tag{6.33}$$

The initial condition is $V_{x=0,\ t=0} = V_0$, which must satisfy the general solution, Eq. (6.32),

$$\text{From } V_0 = (V_0 + b_1), \text{we get } b_1 = 0$$

$$V(x,t) = (V_0 + A_3x + b_2\sin(kx))e^{-k_1^2 t} \tag{6.34}$$

k and k_1 are real constants. Since the electric potential will decrease with distance as the current flows down the axon, A_3 and b_2 are negative real constants.

The membrane potential of the exciting axon may reflect the concentration of positive ions in the axon. The membrane potential of the exciting axon shows a damped oscillation and withers over time and the distance from the origin of the current injection. The diffusion of sodium ions in the axon causes oscillations. As the internal axoplasmic resistance is smaller, k_1 should be larger, and the oscillation damps faster. High friction is a prerequisite for consistent oscillation. Interestingly, similar damped oscillation occurs in the oscillatory release of stored Ca^{2+} in the sarcoplasmic reticulum (SR) through RyR2 channels (Section 5.9). In both cases, the ions move rhythmically along with the concentration gradient.

6.10 Neuronal communication of living cells

Living cells can communicate with each other by means of the gap junctions, or ligand-gated channels. Binding between a ligand and its receptor results from intermolecular forces such as electrostatic forces, which cause the conformational change of the target protein. The binding typically changes the three-dimensional shape of the target protein, which switch on various physiological effects depending on the target protein. Ligands may be hormones like epinephrine (EPI), ACTH, and glucagon, or neurotransmitters such as GABA, glutamate, acetylcholine (Ach), and norepinephrine (NE).

Neurotransmitters are signaling chemicals (chemical messengers) of the nervous system. They are released from synaptic vesicles formed in the axon terminals of a presynaptic neuron, bind selectively to the receptors on the ion channels in the postsynaptic membrane of the target cells such as neurons or skeletal muscle fibers, and the binding changes the conformation of a hydrophilic part of the pore columnar protein and opens the ion-selective channels for Na^+, Cl^-, Ca^{2+}, and K^+ to be able to pass through. ACh at neuromuscular junctions, ACh at autonomic ganglions at the junction of pre- and postganglionic neurons, and others such as glutamate and GABA in the central nervous system are examples of neurotransmitters involved in the transmission of impulses between different cells.

G protein-coupled receptors bind selectively to hormones such as epinephrine, ACTH, and glucagon, or neurotransmitters such as acetylcholine secreted from postganglionic parasympathetic nerves, norepinephrine released from postganglionic sympathetic nerves. The binding to the receptors relays signals down to the cell interior and operate by the subunits of G proteins to induce metabolic responses via second messengers such as cyclic AMP in cardiac and smooth muscle, and IP_3 in smooth muscle [7].

6.10.1 Communication between neurons

Interestingly along with a continuous action of Na^+-K^+ pump, selective permeability of the membrane causes the membrane potential. The voltage-gated ion channels in the membrane of the axon open and generate action potentials when the voltage sensor detects the membrane potential above the threshold, and the action potential propagates down the axon in the way that the action potential is renewed at every voltage-gated Na^+ channel along the axon.

Neurons can communicate with one another through a specialized junction called a synapse. When an action potential from a presynaptic neuron reaches the axon terminals, calcium ions are released from the voltage-gated calcium channels and make vesicles full of small molecules called neurotransmitters fuse with the membrane and dump their contents into the space between the two adjacent neurons called a synaptic cleft. The neurotransmitters such as ACh, glutamate, and gamma-aminobutyric acid (GABA) diffuse across the synaptic cleft and open ion channels on the dendrites or cell body of the postsynaptic neuron to cause postsynaptic potentials. Since the graded potential is the result of a special summation of postsynaptic potentials, it is not always excitatory. Sometimes it can be even inhibitory. If spatial summation is between excitatory and inhibitory postsynaptic potentials, canceling of one another requires more excitatory postsynaptic potentials to generate an action potential, which can be seen in the central nervous system.

6.10.2 Communication between neuron and skeletal muscle fiber

A motor unit consists of a neuron plus the muscle cells it supplies. There are many motor units in a nerve that collectively provide electrical input to a muscle in order to generate a muscle contraction. The axons are insulated with a myelin sheath. The myelin allows the electrochemical signal in the axon to travel over long distances. The upper motor neuron unit originating in the cortex is activated and sends a shower of action potentials to the lower motor neurons in the ventral horn of the spinal cord. From the spinal cord, the motor neuron is activated to generate a current that travels the axon to the neuromuscular junction. Motor and sensory neurons travel together from the nerve root to the limbs. Along the way, sensory nerve twigs split off from the main nerve trunk to the skin to receive sensation while motor nerve twigs split off and supply muscles for contraction. Each skeletal muscle cell can communicate with a motor neuron at one neuromuscular junction (Fig. 6.7). When the neurotransmitter, i.e., ACh discharged from the axon terminals of a motor neuron at a small gap called a synapse binds to the receptor on the ACh-gated sodium ion channels located in the postsynaptic membrane which belongs to the skeletal muscle fiber, the sodium ion channels open. The opening of sodium ion channels causes influx of ions, which leads to the excitatory postsynaptic potential. The membrane except for the area at the neuromuscular junction has voltage-gated Na^+ and K^+ channels. Once an action potential is generated at the membrane close to the junction, the action potential will propagate along the cell membrane of the muscle fiber in both directions and lead to the muscle fiber contraction. Stimulation of a motor neuron will evoke contraction of the innervated muscle fibers. Since muscle fibers have the voltage-gated Na^+ and K^+ channels, the action potential transmits down to the transverse tubules and cause the voltage-sensitive tubule proteins (dihydropyridine receptor) to change shape. This shape change opens the ryanodine type1 Ca^{2+} channels in the sarcoplasmic

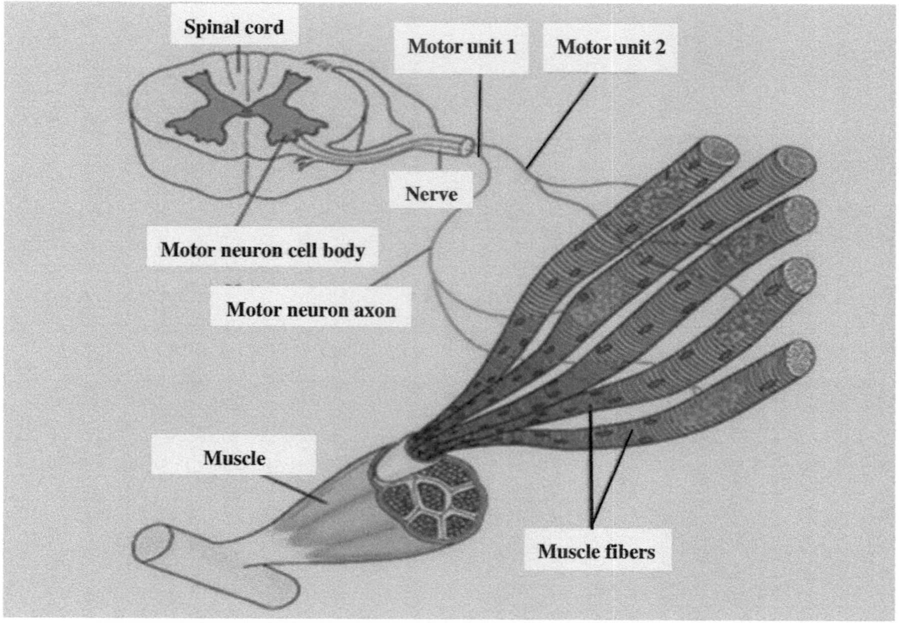

Fig. 6.7 A single somatic motor neuron will innervate multiple muscle fibers. Each skeletal muscle fiber has one neuromuscular junction. A motor unit consists of a neuron plus the muscle fibers it supplies. There are many motor units in a nerve that collectively provide electrical input to a muscle to generate a muscle contraction. From the spinal cord, the motor neuron is activated to generate a current that travels the axon to the neuromuscular junction. Motor and sensory neurons travel together from the nerve root to the limbs. Along the way, sensory nerve twigs split off from the main nerve trunk to the skin to receive sensation while motor nerve twigs split off and supply muscles for contraction. Each skeletal muscle cell can communicate with a motor neuron by a synapse. Stimulation of a motor neuron may evoke contraction of the innervated muscle fibers.

reticulum by means of physical contact, allowing a massive amount of Ca^{2+} to flow into the cytosol and a contraction of myofibrils arises.

6.10.3 Communication between neuron and cardiac or smooth muscle fibers

Cardiac muscle and intestinal smooth muscle have innervations of postganglionic parasympathetic and sympathetic nerves. The postganglionic sympathetic fibers terminating in the SA node and the AV node may play an important role in accelerating the heart rate, and some fibers distributing to the ventricular muscle fiber may increase the force of muscular contractions. Postganglionic parasympathetic fibers supplying the specialized cardiac muscle cells of the SA node and AV node may play an important role in slowing the heart rate, and some fibers distributing to the ventricular muscle fiber may decrease the force of muscular contractions [8]. Cardiac and intestinal muscles have pacemaker cells. The impulse initiated by special pacemaker cells in the SA node propagates directly from cell to cell through the gap junctions and through the conductive system to coordinate rhythmic contractions of the heart muscle. Autonomic nerve impulses cannot initiate the rhythmic contraction of cardiac muscle or the tone of smooth muscle, but they can modulate the rate of contraction via intracellular second messengers and the subunits of G protein acting on inward-rectifying potassium K^+ channels directly as an effector [9]. Various physiological functions are controlled by the autonomic nervous system, mediated by transmitters such as ACh and norepinephrine (NE) binding to G protein-coupled receptors. Human bronchial smooth muscle receives little or no sympathetic innervation, but it contains large numbers of β_2 adrenergic receptors while all smooth muscles of blood vessels have α_1 and β_2 adrenergic receptors and innervations of sympathetic nerve, but no parasympathetic nerve. Calcium ion and cyclic AMP in cardiac and smooth muscles, and inositol triphosphate in smooth muscles are intracellular second messengers (see Chapter 8).

In the cardiac muscle, the release of ACh from the vagus nerve in the parasympathetic notches to pacemaker cells in the SA node stimulates muscarinic M_2 cholinergic receptors (G_i protein-coupled receptors), promotes the dissociation of the α subunit from $\beta\gamma$-complex, and leads to inhibition of adenylyl cyclase, which causes a decrease in cyclic AMP followed by inactivation of voltage-gated Ca^{2+} channels in the plasma membrane. In addition, once the inward-rectifying potassium channel (IKACh) is opened by binding to dissociated $\beta\gamma$-complex, K^+ ions flow out of the pacemaker cells causing hyperpolarization in a resting state [10,11] (see Section 7.3). In its hyperpolarized state, action potentials cannot be generated as quickly, which slows down the heart rate. The decrease in cytosolic Ca^{2+} concentration and increase in the outward current of potassium during diastole may increase the maximum diastolic potential that is the most negative level attained during the cardiac cycle, which accounts for the negative effects of ACh on the heart rate. Stimulation of β_1 adrenergic receptors (G_s protein-coupled receptors) by NE or epinephrine (EPI) leads to an activation of adenylyl cyclase. An increase in cyclic AMP raises the cytosolic Ca^{2+} concentration by activating HCN ion channels [12] and activates the voltage-gated Ca^{2+} channels through activated cyclic AMP-dependent protein kinase (PKA) [13,14] which causes cardiac muscles more excitable (see Section 7.9).

In the intestinal smooth muscle, ACh secreted from postganglionic parasympathetic nerve terminal binds to the muscarinic cholinergic receptors. Muscarinic M_3 receptors are abundant in smooth muscle and glands. The M_3 receptor is a G_q protein-coupled receptor, and its metabolic function is mediated by interactions with G proteins. The dissociated α subunit of G_q proteins are responsible for the stimulation of phospholipase C activity which catalyzes the hydrolysis of one of the membrane phospholipids, phosphatidylinositol 4,5 diphosphate [7]. The hydrolysis produces one of the major second messengers, inositol triphosphate (IP_3) that elicits a massive discharge of Ca^{2+} from intracellular stores in the endoplasmic reticulum through IP_3-gated Ca^{2+} channels. The nerve impulses cannot initiate the tone, instead, they can modulate it. Ca^{2+} binds to calcium-binding protein, calmodulin (CaM), and the Ca^{2+}-CaM then activates calmodulin-dependent myosin light-chain kinase. This activated enzyme catalyzes the phosphorylation of the myosin light chain at serine, which induces activation of ATPase on the actin-binding site of myosin. Cross bridges form between myosin and actin, resulting in the contraction of smooth muscle.

In addition, myosin light-chain kinase from smooth muscle has been known to be phosphorylated by cyclic AMP-dependent protein kinase (PKA), which leads to a decrease in the affinity of the calmodulin-dependent light-chain kinase to Ca^{2+}-CaM and hence, a decrease in the enzymatic activity. This inactivated enzyme cannot catalyze the phosphorylation of the myosin light chain at serine, hence leading to smooth muscle relaxation. This event has been proposed as a mechanism for the relaxation of smooth muscle in response to increased intracellular concentrations of cyclic AMP and can explain the mechanism of the inhibitory response of smooth muscles to stimulation of β_2 adrenergic receptors in the peripheral nervous system [15].

All smooth muscles of blood vessels have innervations of sympathetic nerves. Stimulations of sympathetic nerves on α_1 receptors cause vasoconstriction with an activation of phospholipase C, and stimulations on β_2 receptors cause vasodilatation with an activation of adenylyl cyclase followed by an increase in cyclic AMP and an activation of protein kinase A, which results in phosphorylation of myosin light-chain kinase, hence, decreased affinity of the light-chain kinase to the Ca^{2+}-calmodulin, and the resultant smooth muscle relaxation. Blood vessels of the skin constrict responding stimulation of sympathetic nerves but show no response to β_2 adrenergic receptor agonists, which indicates that blood vessels of the skin have only α_1 receptors.

Human bronchial smooth muscle receives little or no sympathetic innervation, but it contains large numbers of β_2 adrenergic receptors [16]. Stimulation of β_2 adrenergic receptors coupled to Gs proteins leads to the activation of adenylyl cyclase, increased cyclic AMP, and the activation of cyclic AMP-dependent protein kinase (PKA), which results in phosphorylation of myosin light-chain kinase, hence, decreased affinity of the light chain kinase to the Ca^{2+}-calmodulin, and the resultant smooth muscle relaxation [15]. β_2 adrenergic receptor agonists also increase Ca^{2+}-activated K^+ currents in airway smooth muscle, leading to membrane hyperpolarization and relaxation.

6.11 Nerve injuries

Nerve injuries can affect either the sensory nerve, motor nerve, or most commonly both nerves since they travel together. When extensive myelin loss occurs, the electrochemical signal in the axon may fail to conduct all the way down to the muscle, producing distinctive patterns of sensory and motor deficit. The axon can be injured either by a knife wound, compression, infection, diabetes, or alcohol use. There are five locations at which weakness can occur in the peripheral nervous system. They are the motor nerve cell, the nerve root, the neuromuscular junction, and the muscle.

If only the motor nerve cell body is injured and no evidence of a sensory deficit is observed, the site of injury is in the anterior horn of the spinal cord. Because of location, there may be other neurologic signs of spinal cord injury. Examples of this type of injury include infectious, degenerative, neoplastic, genetic, and other conditions. The most common case was poliomyelitis that selectively injured the anterior horn cells in the spinal cord.

The nerve root injuries are commonly due to narrowing of the openings in the side of the spine for nerve root exit. The clinical history often includes severe shooting pain from the spine into the limb and can be accompanied by focal motor, or sensory findings in a specific nerve root distribution.

Wallerian degeneration is an active process of degeneration that results when an axon is cut or crushed and the part of the axon farther from the cell body degenerates. Researchers have believed that axons are highly dependent on their cell bodies for long-term survival. However, recent studies point to the existence of axon-autonomous mechanism(s) that regulate rapid axon degeneration after axotomy. A related process of dying back or retrograde degeneration known as Wallerian-like degeneration occurs in many neurodegenerative diseases, especially those where axonal transport is impaired such as Amyotrophic lateral sclerosis (ALS) and Alzheimer's disease [17]. A failure to deliver enough essential axonal protein, i.e., nicotinamide mononucleotide adenylyltransferase is a key initiating degeneration. Wallerian degeneration occurs after axonal injury in both the peripheral nervous system (PNS) and central nervous system (CNS). It occurs in the section of the axon distal to the site of injury and usually begins within 24–36 h of a lesion. Prior to degeneration, the distal section of the axon tends to remain electrically excitable. After the injury, the axonal skeleton disintegrates, and the axonal membrane breaks apart. Axonal degeneration is followed by degradation of the myelin sheath. The macrophages serve to clear the debris from the degeneration.

Neurapraxia is the least severe form of nerve injury, with complete recovery. In this case, the axon remains intact, but there is myelin damage causing an interruption in the conduction of the impulse down the nerve fiber. Most commonly, this involves compression of the nerve or disruption to the blood supply (*ischemia*). There is a temporary loss of function which is reversible within hours to months of the injury (the average is 6–8 weeks). Wallerian degeneration does not occur. There is frequently greater involvement of motor than sensory function with the autonomic function being retained.

Cerebral infarction causes reduced blood flow usually as a result of a partial or complete blockage of small arterioles. The conduction of impulse will be slowed since disruption to the blood supply damages myelin sheath, and in addition without sufficient oxygen, the electron transfer stops in mitochondria, and the ATP synthesis comes to a halt, and then F_0F_1-ATP synthase works in the reverse direction as proton pump using chemical energy of ATP hydrolysis, and then ATP-deficiency may disturb for Na^+-K^+ ATPases to pump Na^+ out of the cytosol. Symptoms of cerebral infarction are determined by the parts of the brain affected. If the infarct is located in the primary motor cortex, contralateral hemiparesis occurs. Infarction will result in weakness and loss of sensation on the opposite side of the body. Physical examination of the head area will reveal abnormal pupil dilatation, light reaction, and lack of eye movement on the opposite side. If the infarction occurs on the left-side brain, the speech will be slurred.

6.12 Conclusions

The speed of ionic currents depends on the internal resistance, membrane capacitance, and membrane resistance of the cell. When the cell is at rest and most of the ion channels are closed except for some leak ion channels, the cell is said to be in a high resistance state, and the neuron is said to be in a resting state. Efflux of potassium ions in a resting state will cause hyperpolarization. When the graded potential at the axon hillock rises above the threshold, many voltage-gated Na^+ channels open. The cell membrane is said to be in a high conductance state, and the neuron is said to be in an exciting state. The influx of sodium ions at an exciting state will generate an action potential in the axon. The smallest motor neuron with the low membrane capacitance is activated first by an excitatory input while a large motor neuron is recruited later. A larger amount of Na^+ charge stored on the surface of the myelinated membrane with the low capacitance will discharge in an exciting state with both the channel pore and the inactivation gate open, and the amount of discharge will exponentially decrease over time, and the minimum charge on the surface of the membrane will be kept shortly after the gate is closed until repolarization occurs. The cable equation was derived and solved. The membrane potential of the exciting axon shows a damped oscillation and withers over time and distance. High friction is a prerequisite for consistent oscillation. Interestingly, similar damped oscillation occurs in the oscillatory release of stored Ca^{2+} in the SR through RyR2 channels.

When extensive myelin loss occurs, the electrochemical signal in the axon may fail to conduct all the way down to the muscle, producing distinctive patterns of sensory and motor deficit.

References

[1] A. Oshima, Structure and closure of connexin gap junction channels, FEBS Lett. 588 (8) (2014) 1230–1237.
[2] M.E. Del Valle, T. Cobo, J.L. Cobo, J.A. Vega, Mechanosensory neurons, cutaneous mechanoreceptors, and putative mechanoproteins, Microsc. Res. Tech. 75 (8) (2012) 1033–1043.

[3] C.S. Hampe, H. Mitoma, M. Manto, GABA and Glutamate: Their Transmitter Role in theCNS and Pancreatic Islets, IntechOpen, 2018, https://doi.org/10.5772/intechopen.70958 (Chapter 5).
[4] W.A. Catterall, K. Mackie, Local anesthetics, in: The Pharmacological Bases of Therapeutics, eleventh ed., McGraw-Hill, New York, 2006, pp. 369–386.
[5] L.C. Kruger, L.L. Isom, Voltage-gated Na^+ channels: not just for conduction, Cold Spring Harb. Perspect. Biol. (2016) a029264, https://doi.org/10.1101/cshperspect.a029264.
[6] A.M. Krensky, F. Vincenti, W.M. Bennet, Immunosuppressants, tolerogens, and immunostimulants, in: The Pharmacological Basis of Therapeutics, eleventh ed., McGraw-Hill, New York, 2006, pp. 1424–1427.
[7] T.C. Westfall, D.P. Westfall, Neurotransmission: the automatic and somatic motor nervous system, in: The Pharmacological Basis of Therapeutics, eleventh ed., McGraw-Hill, NY, 2006, pp. 137–181.
[8] D.K. Detweiler, Regulation of the heart, in: Best & Taylor's Physiological Basis of Medical Practice, ninth ed., The Williams & Wilkins, Baltimore, MD, 1973, pp. 110–129.
[9] A.M. Brown, A. Yatani, Y. Imoto, G. Kirsch, H. Hamm, J. Codina, R. Mattera, L. Birnbaumer, Direct coupling of G proteins to ionic channels, Cold Spring Harb. Symp. Quant. Biol. 53 (1) (1988) 365–373.
[10] L.G. Hommers, M.J. Lohse, M. Bünemann, Regulation of the inward rectifying properties of G-protein-activated inwardly rectifying K^+ (GIRK) channels by $G_{\beta\gamma}$ subunits, J. Biol. Chem. 278 (2) (2003) 1037–1043.
[11] T.T. Ivanova-Nikolova, E.N. Nikolov, C. Hansen, J.D. Robishaw, Muscarinic K+ channel in the heart: modal regulation by G protein βγ subunits, J. Gen. Physiol. 112 (2) (1998) 199–210.
[12] M. Baruscotti, A. Bucchi, D. DiFrancesco, Physiology and pharmacology of the cardiac pacemaker ("funny") current, Pharmacol. Ther. 107 (1) (2005) 59–79.
[13] W. Osterrieder, et al., Injection of subunits of cyclic AMP-dependent protein kinase into cardiac myocytes modulates Ca2+ current, Nature 298 (1982) 576–578.
[14] B.L. Gerhardstein, T.S. Puri, A.J. Chien, M.M. Hosey, Identification of the sites phosphorylated by cyclic AMP-dependent protein kinase on the β_2 subunit of L-type voltage-dependent calcium channels, Biochemistry 38 (32) (1999) 10361–10370.
[15] J.R. Miller, P.J. Silver, J.T. Stull, The role of myosin light chain kinase phosphorylation in beta-adrenergic relaxation of tracheal smooth muscle, Mol. Pharmacol. 24 (2) (1983) 235–242.
[16] B.J. Undem, Pharmacotherapy of asthma, in: The Pharmacological Basis of Therapeutics, eleventh ed., McGraw-Hill, New York, 2006, pp. 720–721.
[17] M.P. Coleman, M.R. Freeman, Wallerian degeneration, Wld(s), and Nmnat, Annu. Rev. Neurosci. 33 (1) (2010) 245–267.

CHAPTER 7

Electrochemical signaling mechanism in cardiac muscle

7.1 Introduction

Cardiac muscle fibers have a striated appearance. Muscle fibers contain many myofibrils and contract by means of the sliding filament mechanism like skeletal muscle. In contrast to skeletal muscle, cardiac muscle fibers are shorter and interconnected which makes coordinate contractions possible without having individual innervation. The cyclic release of stored Ca^{2+} from the SR (sarcoplasmic reticulum) produces the tone of the muscle and was thought to be a basic source of rhythmic cardiac automaticity. Cardiac muscle has innervation of postganglionic parasympathetic and sympathetic nerves which plays an important role in the modulation of the automaticity.

If electrical stimulations are added to the vagus nerve which is the parasympathetic nerve, the heart slows down beating. The experiment that Dr. Loewe did in 1921 was to connect two frog hearts via the small glass container full of oxygenated Ringer's solution and stimulate the nerve from the head to one heart. The solution perfusing the first heart A was applied to the second heart B via the glass container. When he stimulated the nerve to frog heart A, it slowed the beating, but a little while later heart B also showed the reduced beating. The experiment suggested that there was a substance diffusing from heart A to heart B. The diffusible substance that slowed heart A and heart B beating was called the vagus stuff. We now know that the vagus stuff is ACh. Adding ACh in the same solution also reduces heart beating. It acts on acetylcholine receptors called the muscarinic acetylcholine receptor which is entirely different from the nicotinic acetylcholine receptors on the membrane of skeletal muscle fiber at the neuromuscular junction. These muscarinic acetylcholine receptors are activated by the mushroom toxin, muscarine.

If the released ACh from the stimulated vagus nerve acted on the ACh-gated Na^+ channel in the sarcolemma of the muscle fiber, the postsynaptic depolarization would generate the action potential, causing Ca^{2+} discharge

Modeling Electrochemical Dynamics and Signaling Mechanisms in Excitable Cells with Pathological Case Studies
https://doi.org/10.1016/B978-0-323-98803-2.00001-8

from the sarcoplasmic reticulum with a resultant contraction of the cardiac muscle. However, ACh induces slowing of the heart rate and a negative inotropic effect on hearts, i.e., a decrease in the force of muscular contractions. Today, we know muscarinic acetylcholine receptors in the cardiac muscle are G_i protein-coupled receptors which are called muscarinic M_2 cholinergic receptors in pacemaker cells. The receptors are on the polar region of columnar proteins embedded in the cell membrane to detect ACh outside the cell. When ACh from the vagus nerve binds selectively to the receptor in the cardiac muscle, the conformational change activates dissociation of α subunit from the G_i protein that can inhibit adenylyl cyclase activity and operate negative action by means of a decrease in intracellular second messenger, cyclic AMP [1]. A decrease in cyclic AMP level causes the inhibition of hyperpolarization-activated cyclic nucleotide-gated (HCN) cation channels (see Section 7.6) and a decrease in cytosolic Ca^{2+}, which make the heart less excitable. In addition, muscarinic M_2 cholinergic receptors (G_q protein-coupled receptors) dissociate $\beta\gamma$ subunit to couple to muscarinic potassium channels to increase in outward currents of potassium [2], which cause hyperpolarization in a resting state called maximum diastolic potential and the resultant slowing of the heart rate. The decrease in cytosolic Ca^{2+} and slowing of the heart rate may account for the negative effects of ACh. We will discuss it in detail in this chapter.

On the other hand, stimulation of β_1 adrenergic receptors (Gs protein-coupled) mediates an activation of adenylyl cyclase and a resultant increase in cyclic AMP in the cardiac muscle. Stimulation of sympathetic nerve stimulates β_1 adrenergic receptors, which elicits an increase in cyclic AMP. The cAMP increases cytosolic Ca^{2+} level [3], which is followed by the activation of the Na^+-Ca^{2+} exchanger, resulting in an influx of net cations (3 Na^+ in, 1 Ca^{2+} out) thereby causing cellular depolarization. And together with an inhibition of inwardly rectifying K^+ currents [4] and activation of HCN channels, an increase in Ca^{2+} is implicated in making cardiac muscles to be more excitable. Thus, the sympathetic nervous system can modulate the excitability of the pacemaker cells through the second messenger cAMP. Hence, we call the cAMP the modulator of cardiac automaticity which may increase the frequency of action potential in the sinoatrial (SA) node pacemaker cells.

An action potential initiated at the SA node pacemaker cells propagates to the atrioventricular (AV) node via the internodal pathways, and to ventricular contractile cells via the His-Purkinje cells and causes rhythmic coordinate contraction of the heart muscle. Nerve impulses cannot initiate the rhythmic contraction of cardiac muscle, but they can modulate the contraction rate and its intensity. The membranes of adjacent cardiac muscle fibers interlock each other. The impulse can pass directly from cell to cell through the gap junctions. Thus, heart muscle fibers do rhythmic coordinate contractions without having individual innervation.

7.2 Function of inwardly rectifying K^+ channels in the cardiac myocytes

Inwardly rectifying potassium (K_{ir}, IRK) channels are activated by the membrane phopholipid, i.e., phosphatidylinositol 4,5-bisphosphate (PIP_2) and they pass positive potassium ions more easily in the inward direction into the cell. The inward rectification was discovered in cardiac muscle cells in the 1960s [5] and in 1970 in skeletal muscle cells [6]. In cardiac myocytes, the conductance of IRK channels dominates the resting conductance of leak K^+ channels. IRK channel attenuates outward current below −70 mV essential for rapid restoration of the resting potential from hyperpolarization probably by means of plugging the channel pore. And it restores an outward current between −70 and −55 mV essential for stabilization of the resting membrane potential probably by means of removal of the plug from the channel pore. However, in the operating potential range from −55 to +30 mV, IRK current in the outward direction decreases probably through closing the channel pore again, allowing more potassium ions to stay in the cell and delaying the repolarization (Fig. 7.1). IRK current is thought to be involved in determining the shape of the cardiac action potential through inducing rapid final finish of hyperpolarization, setting the resting potential, and permitting the plateau phase of action potential [7]. In addition, IRK may play an important role in removing small irregular depolarizations during the diastole and protecting the heart from premature contractions (see Section 10.2).

The phenomenon of inward rectification of IRK channels is probably the result of high-affinity block by endogenous polyamines, namely spermine, as well as magnesium ions, that plug the channel pore at both above −55 mV [8] and below −70 mV, resulting in a decrease in outward currents. This voltage-dependent block by polyamines will result in attenuated repolarization.

IRK channels differ from voltage-gated potassium channels since the voltage-gated potassium channels carry outward potassium currents in their operating voltage range and are responsible for repolarizing the exciting cells

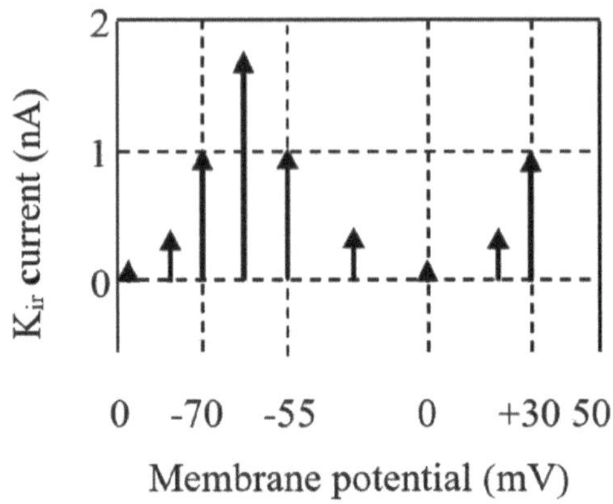

Fig. 7.1 IRK current-membrane potential relationship of cardiac muscle fiber. IRK channel attenuates outward current below −70 mV essential for rapid restoration of the resting potential from hyperpolarization probably by means of plugging the channel pore and increases an outward current between −70 and −55 mV essential for stabilization of the resting membrane potential probably by means of removal of the plug from the channel pore. However, in the operating potential range from −55 to +30 mV, IRK current in the outward direction decreases probably through closing the channel pore again, allowing more potassium ions to stay in the cell and delaying the repolarization.

whereas IRK channels play an important role in setting the resting potential, permitting the plateau phase, inducing rapid final finish of hyperpolarization, and thus preventing the heart from arrhythmias. An increase in cyclic AMP level as seen in sympathomimetic stimulation causes inhibition of inwardly rectifying K^+ currents [4] along with an increase in Ca^{2+} in the myocardium, which may increase an incident of premature contractions (see Section 10.3).

7.3 G protein-coupled receptors in the heart

G protein-coupled receptors (GPCR) are transmembrane proteins that detect ligands outside the cell and activate cellular responses by means of signal transduction pathways. When a ligand binds to the GPCR, it causes a conformational change in the GPCR, which allows it to act as a guanine nucleotide exchange factor. The GPCR activates the coupled G protein by exchanging GDP bound to the G protein for GTP. The G protein's α subunit together with the GTP, then dissociates from the βγ subunits and further affects intracellular target enzymes such as adenylate cyclase or phospholipase C directly depending on the α subunit type.

(1) Signal transduction pathways via G_{α}: (a) cAMP signal pathway is involved in β_1 adrenergic receptors and muscarinic M_2 receptors in cardiac muscle cells, and β_2 adrenergic receptors in the smooth muscles of coronary blood vessels. β_1 and β_2 adrenergic receptors are G_s-coupled receptors. Muscarinic M_2 receptors are G_i-coupled receptors. The effector of dissociated α subunit of G_s and α subunit of G_i is the cAMP-producing enzyme, adenylate cyclase. Stimulation of G_s-coupled receptors activates adenylate cyclase, and stimulation of G_i-coupled receptors inactivate adenylate cyclase. (b) The phosphatidylinositol signal pathway is involved in α_1 adrenergic receptors that is G_q-coupled receptors in coronary blood vessels. The effector of dissociated α subunit of G_q ptotein is the inositol triphosphate-producing enzyme, phospholipase C. The major second messenger, inositol triphosphate (IP_3) diffuses inside the cell, where it binds to its receptor, which is on the IP_3-gated Ca^{2+} channel located in the endoplasmic reticulum (ER) (see Section 8.5).

(2) Signal transduction pathways via $G_{\beta\gamma}$: G protein-gated inward-rectifying potassium (GIRK, IKACh) channel located in the sinoatrial (SA) node and atria in the heart plays a major role in the regulation of heart rate [9]. Activation of the IKACh channels begins with the release of ACh from the vagus nerve to pacemaker cells in the heart. ACh binds to the M_2 muscarinic receptors (G_i-coupled), which promotes the dissociation of $\beta\gamma$-complex from α subunit [10]. The dissociated $\beta\gamma$-complex binds directly and specifically to the IKACh channel through interactions with both the GIRK1 and GIRK4 subunits. (The IKACh channel is composed of two homologous GIRK channel subunits: GIRK1 and GIRK4.) Once the plug in IKACh channel is removed by binding to the $\beta\gamma$-complex, K^+ ions can flow out of the pacemaker cells causing hyperpolarization in a resting state during the diastole [11]. In its hyperpolarized state, action potentials cannot be generated as quickly, which slows the heart rate.

7.4 Ryanodine receptor 2 channels

Ryanodine receptor 2 (RyR2) channel is Ca^{2+}-gated Ca^{2+} channel found primarily in cardiac muscle. The RyR2 channel located in the sarcoplasmic reticulum (SR) is leaky even at rest and supplies Ca^{2+} to the cardiomyocytes to depolarize rhythmically. To enable cardiac muscle contraction, voltage-gated L-type calcium channels in the plasma membrane must be opened by the voltage above the threshold, and the calcium influx must raise an action potential which causes dihydropyridine (DHP) Ca^{2+} channels to open and allows calcium ions to bind to RyR2 located on the SR. This binding causes the massive release of calcium through RyR2 from the SR into the cytosol, where it binds to the C domain of troponin, which uncovers myosin-binding sites on actin causing the formation of the cross-linkages between actin and a myosin head with the high energy phosphate, and the successive conformational change of myosin head through the removal of the phosphate causes sliding of thin filament on the thick filament. The hydrolysis of ATP by the actin activated-myosin ATPase in the myosin head causes the myosin head to bind to the high-energy phosphate in the presence of Ca^{2+} and a recocking of the myosin head to ready for rebinding to actin, which enables the cyclic cardiac muscle contractions (see Section 8.4).

Phosphorylation of RyR2 by activated protein kinase A through sympathomimetic stimulation may make the RyR2 Ca^{2+} channel leakier [12]. In vitro, PKA can phosphorylate RyR2. However, in intact myocytes, a functional role of PKA regulation of RyR2 remains controversial [13]. β-Adrenergic receptor agonist, isoproterenol enhanced Ca^{2+} leak from the SR in intact rabbit ventricular myocytes. The Ca^{2+} leak seemed not to be related to protein kinase A activity but was thought to be mediated by CaMKII activity [14,15]. An increase in the cytosolic Ca^{2+} along with an increase in cAMP will cause cardiac muscle more excitable.

7.5 Sodium-calcium exchanger

The sodium-calcium exchanger (Na^+-Ca^{2+} exchanger or NCX) is a membrane protein that removes Ca^{2+} from the cells. It uses the electrochemical potential energy to allow Na^+ to flow across the plasma membrane in exchange for the countertransport of Ca^{2+}. A single calcium ion is exported for the import of three sodium ions, which results in depolarization. The exchanger exists in many different cell types including cardiac, skeletal, smooth muscle, and neuronal cells. It is considered one of the most important cellular mechanisms for removing Ca^{2+} in a resting state [16], by which the cytoplasmic concentration of calcium ions in the cell is kept low. The exchanger does not bind very tightly to Ca^{2+}, but it can transport the ions rapidly [17]. Therefore, it is useful for ridding the cell of large amounts of Ca^{2+} in a short time, as it is needed for cardiac cells to get in a resting state quickly. Thus, the exchanger likely plays an important role in regaining the cell's normal calcium concentrations [18].

In a resting membrane potential, the Na^+-Ca^{2+} exchanger pumps Ca^{2+} out of cardiac muscle cells with the help of the large extracellular Na^+ concentration. However, during the upstroke of the action potential in a cardiac muscle fiber, there is a large influx of Na^+ ions. This depolarizes the cell and shifts the membrane potential in a positive direction. A large increase in intracellular Na^+ concentration causes the reversal action of the Na^+-Ca^{2+} exchanger, which is the pumping Na^+ ions out of the cell and Ca^{2+} ions into the cell. However, this reversal of the exchanger lasts only momentarily, and the exchanger returns to its forward direction of flow, that is, pumping the Ca^{2+} out of the cell as the cytosolic sodium concentration goes down [19].

The plasma membrane Ca^{2+} ATPase (PMCA) that exports Ca^{2+} from the cell and sarcoendoplasmic reticulum Ca^{2+} ATPase (SERCA) that imports Ca^{2+} into the SR have a much higher affinity with Ca^{2+} but their transport capacity is much lower. The Ca^{2+} ATPases are capable of effectively binding to Ca^{2+} even when their concentrations are quite low. The Na^+-Ca^{2+} exchanger complements Ca^{2+}-ATPases that have a high affinity but low capacitance, and together they are involved in muscle relaxation, maintenance of Ca^{2+} concentration in the SR in cardiac cells.

The activation of the exchanger is also implicated in the cardiac electrical conduction abnormality known as premature contraction [20]. It is thought that intracellular accumulation of Ca^{2+} causes the activation of the Na^+-Ca^{2+} exchanger as seen in sympathomimetic stimulation. The result is the influx of a net positive charge (3 Na^+ in, 1 Ca^{2+} out), thereby causing cellular depolarization. This abnormal depolarization may lead to premature contraction.

7.6 HCN cation channels and modulation of the cardiac rhythmicity

Regarding the SA node depolarization, the channel activated by hyperpolarization has been reported in the plasma membrane [21]. It is a nonselective channel for cations called hyperpolarization-activated cyclic nucleotide-gated (HCN) cation channel. They respond to bind cyclic AMP when the cell is hyperpolarized and depolarize the cell toward the threshold via cation influx called funny currents (I_f). HCN4 is the main isoform expressed in the sinoatrial node (SA node). The current through the activated HCN channels plays a key role in cardiac rhythmicity.

The HCN channels are implicated in the cardiac rhythmicity but do not likely create cyclic depolarizations in the cytosol. The abolishment of HCN channels in the mutant mice caused the failure of some impulse initiations in SA node pacemaker cells. However intermittent firing of SA node pacemaker cells is still preserved and the activation of β adrenergic receptor by isoproterenol restored the regular firing although its rate was slower than that of control mice [22]. The experimental observation suggested that depolarization caused by I_f through the HCN channels was not the basic source of the automaticity of the pacemaker cell activity and there should be a fundamental source of the automaticity elsewhere. Intracellular Ca^{2+} oscillation, i.e., the sinusoidal release of stored Ca^{2+} has been proposed as a primary source of the rhythmical SA node depolarization (see Section 5.9).

The depolarization by the HCN can be achieved in the presence of cAMP. The depolarization begins when conformational change of β_1 adrenergic receptor (G_s protein-coupled receptor) is brought by binding to norepinephrine (NE) or epinephrine (EPI), which allows it to act as a guanine nucleotide exchange factor. Stimulation of the receptor activates the G_s protein by exchanging GDP bound to the G protein for GTP. The G protein's α subunit together with the GTP then dissociates from the βγ subunits and further activates the intracellular cAMP producing enzyme, adenylate cyclase. When the HCN ion channels are under hyperpolarization, they react with the negatively charged cyclic AMP, open for nonselective positive ions, and then cause the inward positive currents (I_f). Simultaneously, a negatively charged cyclic AMP diffuses and then activates cyclic AMP-dependent protein kinase (PKA), and the liberated catalytic subunits of PKA phosphorylate the Ca^{2+} channels to open [23]. The raised depolarization by the activated HCN channels along with cAMP will induce the opening of the L-type voltage-gated Ca^{2+} channels allowing a larger Ca^{2+} influx, which increases the rate of action potentials in the node pacemaker cells.

7.7 Gap junctional conduction

Cardiac muscle fibers are short and interconnected. The membranes of adjacent cardiac muscle fibers interlock each other. The impulse can pass directly from cell to cell through the gap junctions. Bilateral movements of Ca^{2+} through gap junctional channels make cardiac muscle fibers contract coordinately. Phosphorylation of transmembrane protein connexin 43 (Cx43) is necessary to form a pore in the gap junctions for ions to pass through. The phosphorylation occurs at multiple serine residues. Cx43 is phosphorylated soon after synthesis and its phosphorylation site changes according to match the physiological event when it travels along the endoplasmic reticulum (ER) and Golgi to the plasma membrane, or when it functions as a channel in the gap junction membrane. Cx43 are classified into P_0 (nonphosphorylated isoform), P_1, and P_2 (phosphorylated isoform) according to differences in the electrophoretic mobility on sodium dodecyl sulfate-polyacrylamide gel electrophoresis. The mobility shifts probably occur through the changes in the phosphorylation site on Cx43. The phosphorylation at S365 involved in forming the P_1 isoform apparently helps its travel to or within the plasma membrane. Phosphorylation at S325, S328, and/or S330 occurs only in the gap junctional membrane and is necessary to form P_2 isoform that has a pore [24]. Under ischemic conditions, there is a loss of Cx43 at the intercalated disc as the nonphosphorylated form of Cx43 (P_0) moves to the lateral edges of the myocyte and the gap junction becomes less conductive.

The structure of gap junctional channels in the myocardium or the SA or AV nodes are not only modulated by phosphorylation of connexin 43 [25], but also glycosylation, nitrosylation, or ubiquitination [26]. Increased cytosolic Ca^{2+} concentration [27], reduced ATP concentration, and acidification [28] are known to decrease the conduction through gap junctional channels. Reduced conductivity of the gap junction with reduced ATP concentration probably occurs in the myofilament Ca^{2+} sensitization [29] and cardiac ischemia, which may decrease the ion conductivity of the tissue and lead to arrhythmias (see Section 5.2).

7.8 Action potential in the heart

Cardiac muscle has innervation of postganglionic parasympathetic and sympathetic nerves. However, nerve impulses cannot initiate the rhythmic contraction of cardiac muscle, but they can modulate the contraction rate and its intensity. Rhythmic action potentials are generated at the sinoatrial (SA) node pacemaker cells and propagate along the internodal pathways to the atrioventricular (AV) node, and farther to ventricular contractile cells via the His-Purkinje cells, and cause rhythmic coordinate contractions of the heart muscle. Although individual cardiomyocytes are separated by the plasma membrane, the impulse can propagate directly from cell to cell through the gap junctions. They permit the cells to function as if they were a syncytium. There are three types of cardiac cells specialized for impulse initiation, rapid conduction, and contraction. They are "P cells for impulse initiation in the sinoatrial (SA) and atrioventricular (AV) nodes," "the special conductive cells constituting internodal pathways from the SA node to the AV node and a pathway from the SA node to the left atria, the bundle of His, and thin filaments distributing to ventricular contractile cells called Purkinje fibers for rapid conduction," and "contractile myocardial cells for contraction."

Pale oval P cells are located centrally in the SA and AV nodes and so named since they are pale, primitive, and pacemaker cells. These cells are oval or rounded in contrast to the elongated shape of other myocardial cells. The P cells in the SA node initiate rhythmical impulses automatically at a regular speed, 60–80 beats per minute. If the SA node stopped, the P cells in the AV node would take over firing rhythmical impulses at a slightly slower pace.

The postganglionic sympathetic fibers terminating in the SA node and the AV node play an important role in accelerating the heart rate, and some fibers distributing to the ventricular muscle fiber increase the force of muscular contractions. Postganglionic parasympathetic fibers supplying the specialized cardiac muscle cells of the SA node and AV node play an important role in slowing the heart rate, and some fibers distributing to the ventricular muscle fiber decrease the force of muscular contractions [30].

To answer the question of how rhythmical impulses in pacemaker cells are initiated automatically, the first step is to examine the wave pattern of an action potential in the pacemaker cells and compare it with those of the conductive Purkinje cells and ventricular muscle fibers (Fig. 7.2). Action potentials of the nodal pacemaker cells have a blunt curve of depolarization and repolarization, and the conduction rate of impulses is slow. The pacemaker cells are called the slow-response cells. No sharp upstroke of an action potential suggests no sodium ion influx, a large blunt upstroke of the potential probably represents slow Ca^{2+} influx through the sarcolemma, and a slow rate of repolarization suggests slow K^+ efflux. The large sinusoidal wave pattern of action potentials suggests the periodic extracellular calcium influxes. The existence of cyclic depolarizations in the cytosol and its amplification is essential to induce the Ca^{2+} influxes through the voltage-gated Ca^{2+} channels.

The thin filaments called Purkinje fibers distributing to ventricular contractile cells conduct impulses most rapidly, and the bundle of His and contractile myocardial cells also conduct impulses fast. They are called the fast-response cells. The action potential of the Purkinje fibers and ventricular

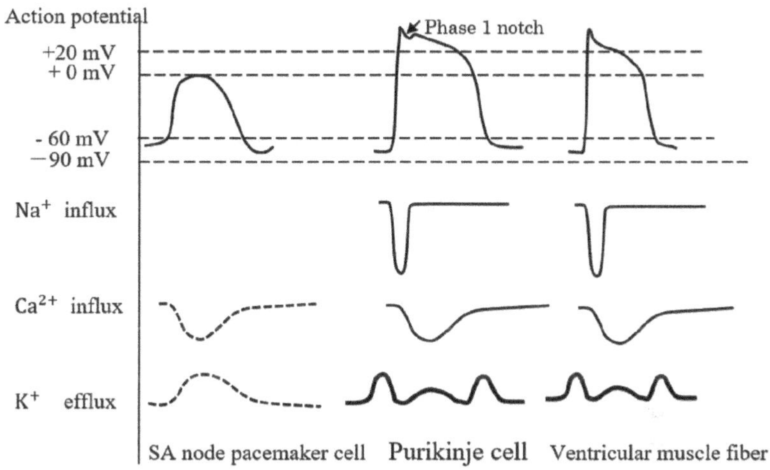

Fig. 7.2 Comparison of the wave pattern of an action potential in the pacemaker cells with those of the conductive Purkinje cells and ventricular muscle fibers. Action potentials of the nodal pacemaker cells have a blunt curve of depolarization and repolarization, and the conduction rate of impulses is slow. No sharp upstroke of an action potential suggests no sodium ion influx, the large blunt upstroke of the potential probably represents slow Ca^{2+} influx, and the slow rate of repolarization suggests slow K^{+} efflux. The large sinusoidal wave pattern of action potentials suggests the periodic extracellular calcium influxes. The sharp initial spike with the phase 1 notch from Purkinje cell may correspond to a rapid depolarization by the opening of voltage-gated Na^{+} channels followed by abrupt closure of the Na^{+} channels. The subsequent prolonged plateau is explained by a slow closure of voltage-gated Ca^{2+} channels and the attenuated K^{+} efflux. IRK channels are dominant in cardiac muscle cells and responsible for the formation of the phase 2 plateau of the action potential through the attenuated outward current of potassium ions in the operating potential range from −55 to +30 mV.

muscle fibers shows an initial sharp upstroke followed by a prolonged plateau of the potential. The sharp initial upstroke may correspond to a rapid depolarization by the opening of voltage-gated Na^{+} channels followed by abrupt closure of the Na^{+} channels. The subsequent prolonged plateau is explained by a slow closure of voltage-gated Ca^{2+} channels and attenuated K^{+} efflux. IRK channels are dominant in cardiac muscle cells and responsible for the formation of the phase 2 plateau of the action potential through the attenuated outward current of potassium ions in the operating potential range from −55 to +30 mV [7] (see Section 7.2). Voltage-gated potassium channels carry outward potassium currents and are responsible for repolarizing the exciting cells. On the other hand, IRK channels are responsible for inducing the rapid final finish of hyperpolarization (see Fig. 7.1) during the diastole.

7.9 Cyclic depolarization of the pacemaker cells and modulation by postganglionic autonomic nerves

Intracellular Ca^{2+} oscillation, i.e., the sinusoidal release of stored Ca^{2+} causes the tone of the muscle. Imaging the SA node cells showed Ca^{2+} sparks arising from the SR [31]. Instead of the HCN channels, the oscillatory release of stored Ca^{2+} from the SR was considered as the basic source of the automaticity of the pacemaker cell activity [32,33]. It was thought to be a diffusion phenomenon and justified mathematically using a model of Ca^{2+} diffusion through the leaky Ca^{2+} channel. Solving the equation of Fick's second law of diffusion, the movement of Ca^{2+} level in the channel showed damped oscillation along the straight line with a negative slope (see Section 5.9).

Using the electrochemical potential energy of Na^+ across the plasma membrane, the sodium-calcium exchanger removes Ca^{2+} from the cells during the diastole. A single calcium ion is exported for the import of three sodium ions, which likely results in amplifying depolarization caused by the oscillatory release of Ca^{2+} from the SR.

Ca^{2+}-activated K^+ channels were found to be a critical player in continuous rhythmic pacemaker activity derived from human embryonic stem cells. [34]. During the last phase of the systolic depolarization, enough cytosolic Ca^{2+} level may activate Ca^{2+}-activated outward K^+ currents which contribute to hyperpolarization. Then hyperpolarization activates HCN channels and raises I_f if there is enough cyclic AMP in the cell. IRK channels work for inducing a rapid final finish of the hyperpolarization and setting the resting potential to get ready for the next depolarization.

The rate of oscillatory depolarization in the SA node pacemaker cells is modulated by postganglionic parasympathetic and sympathetic nerves.

(1) *Slowing rhythmic impulses*: Activation of the IKACh channels begins with a release of ACh from the vagus nerve supplying the specialized myocardial tissue of the SA and AV nodes to pacemaker cells. ACh binds to the M_2 muscarinic receptors (G_i-coupled receptors), which promotes the dissociation of the $\beta\gamma$-complex from the α subunit [10] (see Section 7.3). Once the plug of the IKACh channel is removed by binding to the $\beta\gamma$-complex, more K^+ ions flow out of the pacemaker cells causing hyperpolarization in a resting state during the diastole [11]. In its hyperpolarized state, action potentials cannot be generated as quickly, which slows the heart rate. On the other hand, the dissociated α subunit can inhibit adenylate cyclase activity (Fig. 7.3) and operate negative action by means of a decrease in intracellular second messenger, cyclic AMP. A decrease in cyclic AMP level causes the inactivation of HCN channels, which may slow the heart rate and a reduction of cytosolic Ca^{2+}.

(2) *Accelerating rhythmic impulses*: Activation of the HCN channels begins with a release of NE (norepinephrine) from the postganglionic sympathetic nerve terminating in the SA and AV nodes to pacemaker cells in the heart. NE binds to the β_1 adrenergic receptors (G_s-coupled), which promotes the dissociation of the α subunit. The effector of the α subunit of G_s is the cAMP-producing enzyme, adenyl cyclase (Fig. 7.3). When HCN ion channel is activated by hyperpolarization in the presence of a negatively charged cAMP, it opens for nonselective positive ions, and then causes the inward positive currents (I_f). The depolarization evoked by the HCN channels may cause a large Ca^{2+} influx through the L-type voltage-gated Ca^{2+} channel, which may raise action potentials in the node pacemaker cells. Simultaneously, a negatively charged cyclic AMP diffuses, activates cyclic AMP-dependent protein kinase (PKA), and liberated catalytic subunits of PKA phosphorylate the Ca^{2+} channels to open [3,23]. Through inactivation of inwardly rectifying K^+ currents [4], an increase in Ca^{2+}, and activation of HCN channels, an increase in cAMP by stimulation of the sympathetic nerve is implicated in making cardiac muscles to be excitable.

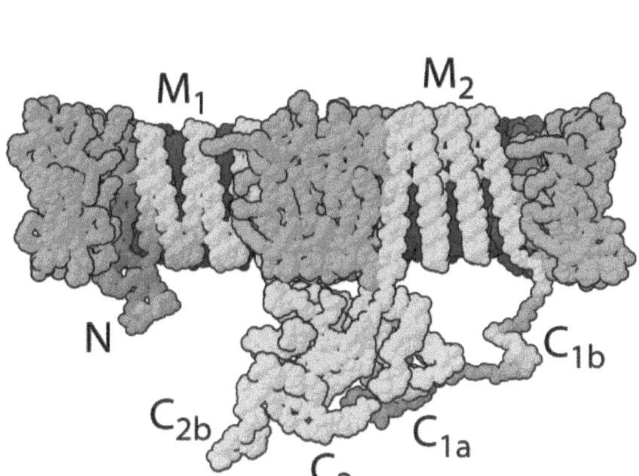

Fig. 7.3 Most adenyl cyclases are transmembrane proteins with 12 transmembrane segments and 2 cytoplasmic domains. The α subunit of G proteins associates with adenyl cyclase, which converts ATP to cAMP.

7.10 Amplification and stabilization of signals

The amplification of Ca^{2+} oscillation from the SR, rapid final finish of the hyperpolarization, setting the stable resting potential, and permitting the plateau phase of an action potential are all essential for cardiac cells to generate stable rhythmic action potentials continuously. The ectopic depolarizations probably are caused by amplified Ca^{2+} in the excitable cardiac cells with reduced conductivity (see Section 10.3).

The cardiac ryanodine (RyR2) channels are Ca^{2+}-gated Ca^{2+} channels in the SR. When leak RyR2 channels are at a resting state, the channel may allow the slow sinusoidal release of stored Ca^{2+} from the SR [35,36] to the cytosol, and then the oscillatory depolarizations may be amplified during the diastole by means of Na^{+}-Ca^{2+} exchanger (see Section 7.5). At the same time, the Ca^{2+} removal by Na^{+}-Ca^{2+} exchanger contributes to keeping continuous pacemaker activity in the SA node [37]. During the last phase of the systolic depolarization, enough cytosolic Ca^{2+} level may activate Ca^{2+}-activated outward K^{+} currents [38] which may bring hyperpolarization, and then the hyperpolarization activates HCN channels to induce the hyperpolarization-activated funny currents (I_f) [21] in the presence of enough cyclic AMP. The amplified depolarizations by Na^{+}-Ca^{2+} exchanger and I_f activates L-type Ca^{2+} channels to discharge large amounts of Ca^{2+} into the cardiac pacemaker cells.

The outward K^{+} currents through inwardly rectifying potassium (IRK) channels during the early phase of depolarization contribute to setting the stable resting potential by removing irregular small noises during the diastole. On the other hand, the attenuated outward K^{+} currents in the operating potential range stabilize the action potential by permitting the plateau phase [7] (see Section 7.2). An increase in cyclic AMP level causes increases in Ca^{2+} and I_f, and inactivation of inwardly rectifying K^{+} channels in the myocardium, which may increase an incident of arrhythmia [4,39].

7.11 Conduction between the atrium and ventricle

The thin filaments called Purkinje fibers distributing to ventricular contractile cells conduct impulses the fastest since they are thin filaments having the most voltage-gated Na^{+} channels in the sarcolemma-like neurons. The velocity of conduction is 4 m/s. The rate of conduction in the bundle of His and contractile myocardial cells is not as fast as the Purkinje fibers, and their velocities are about 1 m/s. The cells which have many voltage-gated

Na^+ channels conduct impulses fast and respond to stimuli quickly, but the refractory period is longer than slow-response cells due to the inactivation of voltage-gated sodium channels.

The conduction rate of the atrioventricular (AV) node is the slowest (0.05 m/s) since pacemaker cells in the AV node lack voltage-gated Na^+ channels where diffusion of Ca^{2+} through gap junctional channels mainly contribute to the propagation of impulses. The refractory period is shorter than fast-response cells. The slow conduction in the AV node is a prerequisite to give the atrial contractile muscle fibers a time to propel all the blood into the ventricle before the ventricles contract and protect the ventricles from supraventricular tachycardia (see Section 10.4.3). The AV node has automaticity (40–60 beats per min) with innervation of postganglionic autonomic nerve fibers. The slow conduction in the AV node can maintain the pulse rate of the ventricles in the physiological range in the case of the atrium fibrillation where the pulse in the atrium is disorganized (see Section 10.2).

7.12 Conclusions

IRK channels differ from voltage-gated potassium channels since the voltage-gated potassium channels carry outward potassium currents in their operating voltage range and are responsible for repolarizing the exciting cells whereas IRK channels play an important role in setting the resting potential, permitting the plateau phase, inducing rapid final finish of hyperpolarization, and thus preventing the heart from arrhythmias. Using the electrochemical potential energy of Na^+ across the plasma membrane, the sodium-calcium exchanger removes Ca^{2+} from the cells during the diastole. A single calcium ion is exported for the import of three sodium ions, which likely results in amplifying depolarization caused by the oscillatory release of Ca^{2+} from the SR.

The cyclic release of stored Ca^{2+} from the SR or the ER produces the tone of the muscle and was thought to be a basic source of rhythmic cardiac automaticity. Sympathomimetic stimulations activate adenylate cyclase to produce the second messenger, cAMP which activates the HCN channels to raise the pulse rate. Through an inactivation of inwardly rectifying K^+ currents [4], an increase in Ca^{2+}, and an activation of HCN channels, an increase in cAMP by stimulation of the sympathetic nerve is implicated in making cardiac muscles to be more excitable.

On the other hand, activation of the IKACh channels begins with a release of ACh from the vagus nerve supplying the SA and AV nodes.

ACh binds to the M_2 muscarinic receptors (G_i-coupled), which promotes the dissociation of the βγ-complex from the α subunit [10]. Once the plug in the IKACh channel is removed by binding to the βγ-complex, K^+ ions flow out of the pacemaker cells causing hyperpolarization in a resting state during the diastole [11]. In its hyperpolarized state, action potentials cannot be generated as quickly, which slows the heart rate. The dissociated α subunit inhibits adenylate cyclase activity and operates negative action by means of a decrease in intracellular second messenger, cyclic AMP.

References

[1] T.C. Westfall, D.P. Westfall, Neurotransmission: the automatic and somatic motor nervous systems, in: The Pharmacological Basis of Therapeutics, eleventh ed., McGraw-Hill, New York, 2006 (Chapter 6).
[2] K. Wickman, G. Krapivinsky, S. Corey, M. Kennedy, J. Nemec, I. Medina, D.E. Clapham, Structure, G protein activation, and functional relevance of the cardiac G protein-gated K^+ channel, I_{KACh}, Ann. N. Y. Acad. Sci. 868 (1) (1999) 386–398.
[3] W. Osterrieder, et al., Injection of subunits of cyclic AMP-dependent protein kinase into cardiac myocytes modulates Ca^{2+} current, Nature 298 (1982) 576–578.
[4] S. Koumi, C.L. Backer, C.E. Arentzen, R. Sato, Beta-adrenergic modulation of the inwardly rectifying potassium channel in isolated human ventricular myocytes. Alteration in channel response to beta-adrenergic stimulation in failing human hearts, J. Clin. Invest. 96 (6) (1995) 2870–2881.
[5] O.F. Hutter, D. Noble, Rectifying properties of heart muscle, Nature 188 (1960) 495.
[6] R.H. Adrian, W.K. Chandler, A.L. Hodgkin, Slow changes in potassium permeability in skeletal muscle, J. Phys. 208 (3) (1970) 645–668.
[7] H. Hibino, A. Inanobe, K. Furutani, S. Murakami, et al., Inwardly rectifying potassium channels: their structure, function, and physiological roles, Physiol. Rev. 90 (2010) 291–366.
[8] A.N. Lopatin, E.N. Makhina, C.G. Nichols, The mechanism of inward rectification of potassium channels: long-pore plugging by cytoplasmic polyamines, J. Gen. Physiol. 106 (5) (1995) 923–955.
[9] E.N. Nikolov, T.T. Ivanova-Nikolova, Coordination of membrane excitability through a GIRK1 signaling complex in the atria, J. Biol. Chem. 279 (22) (2004) 23630–23636.
[10] T.T. Ivanova-Nikolova, E.N. Nikolov, C. Hansen, J.D. Robishaw, Muscarinic K+ channel in the heart: modal regulation by G protein βγ subunits, J. Gen. Physiol. 112 (2) (1998) 199–210.
[11] L.G. Hommers, M.J. Lohse, M. Bünemann, Regulation of the inward rectifying properties of G-protein-activated inwardly rectifying K^+ (GIRK) channels by $G_{\beta\gamma}$ subunits, J. Biol. Chem. 278 (2) (2003) 1037–1043.
[12] S.O. Marx, S. Reiken, Y. Hisamatsu, T. Jayaraman, D. Burkhoff, N. Rosemblit, et al., PKA phosphorylation dissociates FKBP12.6 from the calcium release channel (ryanodine receptor): defective regulation in failing hearts, Cell 101 (2000) 365–376.
[13] D.M. Johnson, G. Antoons, Arrhythmogenic mechanisms in heart failure: linking β-adrenergic stimulation, stretch, and calcium, Front. Physiol. 9 (2018) 1453.

[14] J. Curran, M.J. Hinton, E. Ríos, D.M. Bers, T.R. Shannon, β-adrenergic enhancement of sarcoplasmic reticulum calcium leak in cardiac myocytes is mediated by calcium/calmodulin-dependent protein kinase, Circ. Res. 100 (2007) 391–398.
[15] X. Ai, J. Curran, T.R. Shannon, D.M. Bers, S.M. Pogwizd, Ca^{2+}/calmodulin-dependent protein kinase modulates cardiac ryanodine receptor phosphorylation and sarcoplasmic reticulum Ca^{2+} leak in heart failure, Circ. Res. 97 (12) (2005) 1314–1322.
[16] R. DiPolo, L. Beaugé, Sodium/calcium exchanger: influence of metabolic regulation on ion carrier interactions, Physiol. Rev. 86 (1) (2006) 155–203.
[17] E. Carafoli, L. Santella, D. Branca, M. Brini, Generation, control, and processing of cellular calcium signals, Crit. Rev. Biochem. Mol. Biol. 36 (2) (2001) 107–260.
[18] L. Kiedrowski, G. Brooker, E. Costa, J.T. Wroblewski, Glutamate impairs neuronal calcium extrusion while reducing sodium gradient, Neuron 12 (2) (1994) 295–300.
[19] D.M. Bers, Cardiac excitation-contraction coupling, Nature 415 (6868) (2002) 198–205.
[20] L. Lilly, Pathophysiology of Heart Disease, Mechanisms of Cardiac Arrhythmias, sixth ed., Lippencott, Williams and Wilkens, New York, 2007 (Chapter 11).
[21] M. Baruscotti, A. Bucchi, D. DiFrancesco, Physiology and pharmacology of the cardiac pacemaker ("funny") current, Pharmacol. Ther. 107 (1) (2005) 59–79.
[22] J. Alig, et al., Control of heart rate by cAMP sensitivity of HCN channels, PNAS 106 (29) (2009) 12189–12194, https://doi.org/10.1073/pnas.0810332106.
[23] B.L. Gerhardstein, T.S. Puri, A.J. Chien, M.M. Hosey, Identification of the sites phosphorylated by cyclic AMP-dependent protein kinase on the β_2 subunit of L-type voltage-dependent calcium channels, Biochemistry 38 (32) (1999) 10361–10370.
[24] J.L. Solan, P.D. Lampe, Key connexin 43 phosphorylation events regulate the gap junction life cycle, J. Membr. Biol. 217 (1–3) (2007) 35–41.
[25] J.L. Solan, P.D. Lampe, Connexin 43 phosphorylation-structural changes and biological effects, Biochem. J. 419 (2) (2009) 261–272.
[26] S.R. Johnstone, M. Billaud, A.W. Lohman, E.P. Taddeo, B.E. Isokson, Post-translational modifications in connexins and pannexins, J. Membr. Biol. 245 (2010) 319–322.
[27] B. Rose, I. Simpson, W.R. Loewenstein, Calcium ion produces graded changes in permeability of membrane channels in cell junction, Nature 267 (1977) 625–627.
[28] D.C. Spray, A.L. Harris, M.V.L. Bennett, Gap junctional conductance is a simple and sensitive function of intracellular pH, Science 211 (1981) 712–715.
[29] S. Huke, R. Venkataraman, M. Faggioni, S. Bennuri, H.S. Hwang, F. Baudenbacher, B.C. Knollmann, Focal energy deprivation underlies arrhythmia susceptibility in mice with calcium-sensitized myofilaments, Circ. Res. 112 (10) (2013) 1334–1344.
[30] D.K. Detweiler, Regulation of the heart, in: Best & Taylor's Physiological Basis of Medical Practice, ninth ed., The Williams & Wilkins, Baltimore, MD, 1973, pp. 110–129.
[31] J. Huser, L.A. Blatter, S.L. Lipsius, Intracellular Ca^{2+} release contributes to automaticity in cat atrial pacemaker cells, J. Physiol. 524 (2000) 415–422.
[32] T.M. Vinogradova, V.A. Maltsev, K.Y. Bogdanov, et al., Rhythmic Ca^{2+} oscillations drive sinoatrial nodal cell pacemaker function to make the heart tick, Ann. N. Y. Acad. Sci. 1047 (2005) 138–156.
[33] V.A. Maltsev, E.G. Lakatta, Normal heart rhythm is initiated and regulated by an intracellular calcium clock within pacemaker cells, Heart Lung Circ. 16 (2007) 335–348.
[34] D. Weisbrod, et al., SK4 Ca2+ activated K+ channel is a critical player in cardiac pacemaker derived from human embryonic stem cells, PNAS 110 (18) (2013) E1685–E1694, https://doi.org/10.1073/pnas.1221022110.

[35] T. Watanabe, Biophysical Basis of Physiology and Calcium Signaling Mechanism in Cardiac and Smooth Muscle, Academic Press, San Diego, CA, 2018 (Chapter 1).
[36] R. Zucchi, S. Ronca-Testoni, The sarcoplasmic reticulum Ca^{2+} channel/ryanodine receptor: modulation by endogenous effectors, drugs and disease states, Pharmacol. Rev. 49 (1) (1997) 1–52.
[37] S. Groenke, E.D. Larson, S. Alber, et al., Complete atrial-specific knockout of sodium-calcium exchange eliminates sinoatrial node pacemaker activity, PLoS One 8 (2013) e81633.
[38] D. Weisbrod, A. Peretz, A. Ziskind, N. Menaker, S. Oz, L. Barad, S. Eliyahu, J. Itskovitz-Eldor, N. Dascal, D. Khananshvili, O. Binah, B. Attali, SK4 Ca^{2+} activated K^{+} channel is a critical player in cardiac pacemaker derived from human embryonic stem cells, PNAS 110 (18) (2013) E1685–E1694, https://doi.org/10.1073/pnas.1221022110.
[39] R.C. Myles, L. Wang, D.M. Bers, C.M. Ripplinger, Decreased inward rectifying K^{+} current and increased ryanodine receptor sensitivity synergistically contribute to sustained focal arrhythmia in the intact rabbit heart, J. Physiol. 593 (2015) 1479–1493.

CHAPTER 8

Mechanism of muscle contraction and disorder

8.1 Introduction

Skeletal muscle fiber has striated appearance. The muscle fibers contain many myofibrils. The nerve cells that activate skeletal muscle fibers are called somatic motor neurons or motor neurons of voluntary nervous system. The axon of each motor neuron divides as it enters the muscle, and each ending gives off several short branches called axon terminals that collectively form a neuromuscular junction. A single somatic motor neuron will innervate multiple muscle fibers. Each skeletal muscle fiber has one neuromuscular junction. A motor unit consists of a neuron plus the muscle fibers it supplies. There are many motor units in a nerve that collectively provide electrical input to a muscle in order to make the muscle contract (see Section 6.10.2). The upper motor neuron unit originating in the cortex is activated and sends a shower of action potentials to the lower motor neurons in the ventral horn of the spinal cord. From the spinal cord, the lower motor neuron is activated to generate a current that travels through the axon to the neuromuscular junction. Motor and sensory neurons travel together from the nerve root to the limbs. Along the way, a sensory nerve twigs split off from the main nerve trunk to the skin to receive sensation while a motor nerve twigs split off and supply muscles for contraction. Each skeletal muscle cell can communicate with a motor neuron at one neuromuscular junction. When the neurotransmitter, i.e., ACh discharged from the axon terminals of a motor neuron at a small gap called a synapse binds to the receptor on the ACh-gated sodium ion channels located in the postsynaptic membrane, which belongs to the skeletal muscle fiber, the sodium ion channels open. The sodium ion influx causes the excitatory postsynaptic potential. The membrane, except the area at the neuromuscular junction, has voltage-gated Na^+ and K^+ channels. Once an action potential is generated at the membrane close to the junction, the action potential will propagate along the sarcolemma of the muscle fiber in both directions and lead to the muscle fiber contraction.

Modeling Electrochemical Dynamics and Signaling Mechanisms in Excitable Cells with Pathological Case Studies
https://doi.org/10.1016/B978-0-323-98803-2.00008-0

Cardiac muscle fibers have striated appearance. Muscle fibers contain many myofibrils and contract by means of the sliding filament mechanism like skeletal muscle. In contrast to skeletal muscle, cardiac muscle fibers are shorter and interconnected. The postganglionic parasympathetic and sympathetic nerves terminating in the SA and AV nodes contributes to the modulation of cardiac rhythmicity. Some sympathetic fibers distributing to the ventricular muscle can increase the force of muscular contractions. Since the membranes of adjacent cardiac muscle fibers interlock each other, an electrochemical signal can pass directly from cell to cell through gap junctions. Thus, a heart muscle fiber does not need to have an individual innervation. Action potentials originate in the pacemaker cells of the sinoatrial (SA) node and propagate to the atrioventricular (AV) node via the internodal pathways and spread to ventricular contractile cells via the His-Purkinje cells, and cause coordinate contractions of the heart muscle. The nerve impulses cannot initiate rhythmic contractions of the cardiac muscle, but they can modulate the rate of contractions and its intensity. Through the inhibition of inwardly rectifying K^+ currents and an activation of HCN channels, an increase in cAMP by stimulation of sympathetic nerve is implicated in causing cardiac muscles to be more excitable (see Section 7.9).

Smooth muscle fibers are spindle-shaped cells with one centrally located nucleus and no externally visible striations. It is found mainly in the walls of hollow organs. Smooth muscles of hollow organs such as trachea, bronchi, stomach, intestine, some blood vessels, uterus, urinary bladder, and ureter have innervations of both postganglionic parasympathetic and sympathetic nerves. *The single unit smooth muscles such as stomach, intestine, ureter, small blood vessels, and uterus* have many gap junctional connections between individual cells and act as one sheet in a coordinated fashion. Thus, the muscle fiber does not need to have an individual innervation, and a depolarization of one fiber triggers the synchronous depolarization throughout the bundle, which is independent of its nerve supply and causes continuous partial contractions called the muscle tone [1]. Nerve impulses cannot initiate the tone, instead they can modulate its intensity. The outstanding characteristic of the intestinal muscle is its rhythmic segmental and peristaltic contractions. Smooth muscles of blood vessels have innervations of sympathetic nerve but no parasympathetic innervations except the vessels in skeletal muscle and salivary glands. Stimulations of sympathetic nerves on α_1 receptors cause vasoconstriction, and stimulations on β_2 receptors cause vasodilatation. Blood vessels of skin constrict responding stimulation of sympathetic nerves but show no response to β_2 adrenergic receptor agonists, which indicates that blood vessels of skin have only α_1 receptors. *Smooth muscles of trachea, bronchi, and in the wall of large arteries are made of multi-unit fibers.* The fibers are not connected by gap junctions. Each fiber is innervated like a skeletal muscle fiber, and a depolarization of one fiber is followed by a contraction of only that fiber. Human bronchial smooth muscle has parasympathetic innervation, but little or no sympathetic innervation. However, it has many β_2 adrenergic receptors responding to a circulating hormone, i.e., epinephrine.

8.2 Contraction of skeletal muscle and neuromuscular junction disorder

The nerve cells that activate skeletal muscle fibers are called somatic motor neurons or motor neurons of voluntary nervous system. A single somatic motor neuron may innervate many muscle fibers. Each skeletal muscle fiber has one neuromuscular junction. Thus, the stimulation of a motor neuron evokes a contraction of the innervated muscle fibers. The sarcolemma (membrane of the striated muscle fiber) at the junction may have millions of ACh receptors called the nicotinic acetylcholine receptor, which are activated by nicotine. The binding of the neurotransmitter, i.e., ACh to the receptors on the ACh-gated sodium ion channel opens the sodium ion channels in the sarcolemma. The influx of the ions causes a postsynaptic depolarization. Poor absorption and low lipophilicity of ACh due to a positive charge on quaternary ammonium might cause prolonged effects if they were not quickly broken down by the enzyme, i.e., acetylcholinesterase in the synaptic cleft. The secreted ACh must be cleared away quickly by the enzyme in order to respond immediately to a sequence of frequent stimuli. The sarcolemma has many voltage-gated Na^+ and K^+ channels like a neuron. An action potential is generated at the membrane close to the junction, and the signal will propagate along the sarcolemma in both directions swiftly down to the transverse tubules and cause the voltage-sensitive tubule proteins (dihydropyridine receptor) to change the shape. This shape change opens the ryanodine type 1 Ca^{2+} channels in the sarcoplasmic reticulum by means of the physical contact, allowing a massive amount of Ca^{2+} to discharge into the cytosol and myofibrils to contract. The striated muscle contraction is the result of a reversible interaction of thick myosin filaments with the thin filaments that are composed primarily of actin, troponin, and tropomyosin.

Neuromuscular junction disorder will produce weakness of the muscles because there is a failure of the electrochemical signal conduction from the nerve to the muscle. The classic example is myasthenia gravis. Myasthenia gravis is a serious disease that causes muscle weakness. Patients typically present with symmetric proximal arm or leg weakness associated with normal reflexes and sensation. These can occur from diverse etiologies including inflammatory, genetic, metabolic, drug-induced, and many other conditions associated with the formation of circulating antibodies to the nicotinic cholinergic receptors in the skeletal muscle. These antibodies destroy some of the receptors on the sodium ion channels and disrupt neuromuscular transmission. The major structural abnormality is the appearance of sparse, shallow, and abnormally wide synaptic clefts at the neuromuscular junction. The postsynaptic membrane has a reduced response to ACh. What kinds of treatment have been done in the patients of myasthenia? The administration of AChE reversible inhibitors such as neostigmine or edrophonium improves muscle weakness. Immunosuppressive drug such as prednisone can suppress antibody production and have been shown to improve muscle strength in some patients.

8.3 Joint pain and weakness of the muscle

The most common types of arthritis, osteoarthritis arthritis, is the swelling and tenderness of one or more of joints associated with weakness of the muscle that supports the joint. The main symptoms of arthritis are joint pain and stiffness, which typically worsen with age. Osteoarthritis causes inflammation of the muscle and tendon, and broken cartilage (the hard slippery tissue that covers the ends of bones where they form a joint). Frequency of incidence rises by age, obesity, and weakness of the muscle. An early sign of the initial symptoms of osteoarthritis is joint pain and stiffness caused by the inflammation of the muscle and tendon when starting the first step of walking. The pain is reduced temporarily by means of warming the muscle to improve the circulation. Its treatments are to reduce weights on the joint, improve inflammation by antiinflammatory drugs, and strengthen the muscle that supports the joint.

8.4 Rhythmic contractions of cardiac muscle

When rhythmic deporalizations of the membrane potential in pacemaker cells raises above the threshold (-50 to $-60\,\mathrm{mV}$), plasma membrane

L-type Ca^{2+} channels with voltage-sensor open and generate action potentials. In the heart, an increase in cyclic AMP caused by the sympathtic stimulation to β_1 receptor will induce the excitability of the cardiac cells by means of activation of HCN channels and the voltage-gated calcium ion channels. On the other hand, ACh released from the vagus nerve binds to the M_2 muscarinic acetylcholine receptors (G_i protein-coupled), which promotes the dissociation of the $\beta\gamma$-complex from α subunit. Once the plug in IKACh channel is removed by binding to the $\beta\gamma$-complex, more K^+ ions flow out of the pacemaker cells causing hyperpolarization in a resting state during the diastole (see Section 7.9) and slow the heat rate. The action potential originates in the pacemaker cells in the SA node and propagates from the SA node to the AV node, and spreads to the ventricular muscle by means of the cardiac conduction system that consists of the bundle of His and its thin filaments called Purikinje cells. The depolarization of cardiac cells will induce Ca^{2+} influx through plasma membrane L-type Ca^{2+} channels, generate an action potential, and lead to a contraction of cardiac muscle fibers. The transmission of action potential along the T tubule membrane of the cardiac muscle fiber activates votage-gated dihydropyridine receptors (DHPRs). The opening of dihydropyridine (DHP) Ca^{2+} channels in the T tubule membrane triggers a massive discharge of stored calcium from the SR by activating the ligand-gated ryanodine receptors (RyR2) that is close to the T tubule [2,3]. The ryanodine channels are Ca^{2+}-gated Ca^{2+} channels in the SR, which is the calcium-induced calcium release system. Increase in the cytosolic Ca^{2+} by sympathomimetic stimulation or digoxin administration leads to increased storage of calcium in the SR, thus causing a corresponding increase in an amount of calcium discharge through each action potential, leading to the increased force in a muscle contraction.

The striated muscle contraction is the result of a periodic interaction of thick myosin filaments with the thin filaments that is composed primarily of actin, troponin, and tropomyosin. Myosin binding sites on actin is blocked by tropomyosin during relaxation. Binding of Ca^{2+} from the SR to the C domain of troponin uncovers myosin binding sites on actin, which allows formation of cross-bridge between actin and myosin head with a high-energy phosphate, and the succesive sliding of thin filament on the thick filament occurs as the head bents releasing the phosphate, and myosin ATPase in the head is activated through binding to the actin. Myosin head detaches as ATP attaches to myosin. The hydrolysis of ATP by the activated myosin ATPase in the presence of Ca^{2+} causes myosin head to bind to a high-energy phosphate and a recocking of the myosin head to ready for rebinding to

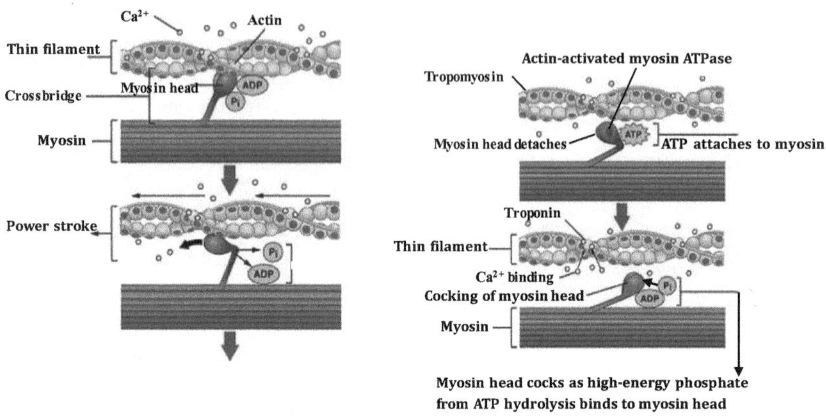

Fig. 8.1 The striated muscle contraction is the result of a periodic interaction of thick myosin filaments with thin filaments that is composed primarily of actin, troponin, and tropomyosin. Myosin binding sites on actin is blocked by tropomyosin during relaxation. Binding of Ca^{2+} from the SR to the C domain of troponin uncovers myosin binding sites on actin, which allows formation of cross-bridge between actin and myosin head with a high-energy phosphate, and the succesive sliding of thin filament on the thick filament occurs as the head bents releasing high-energy phosphate, and myosin ATPase in the head is activated through binding to actin. Myosin head detaches as ATP attaches to myosin. The hydrolysis of ATP by the activated myosin ATPase in the presence of Ca^{2+} causes myosin head to bind to a high-energy phosphate and a recocking of the myosin head to ready for rebinding to actin, which enables the cyclic cardiac muscle contractions.

actin, which enables the cyclic cardiac muscle contractions (Fig. 8.1). The contraction is driven reversibly by the hydrolysis of ATP by myosin ATPase activated by Ca^{2+} and actin. When the SR lowers the Ca^{2+} concentration, relaxation of cardiac muscle occurs since tropomyosin physically blocks the binding of the myosin head to actin and does not allow the myosin ATPase to bind to actin to be activated.

The inhibition of the myosin ATPase activity occurs under low Ca^{2+} level without blocking the binding of myosin to actin [4]. The studies of the regulation of cardiac muscle contraction have suggested that troponin T not only anchors the troponin I-troponin C complex to the thin filaments, but also probably confers the Ca^{2+} sensitivity of the actin-activated myosin ATPase and/or force [5]. If the ATPase in the myosin head becomes more active at lower concentration of Ca^{2+}, the muscle fibers may get more excitable, and may induce arrhythmias. The Ca^{2+} overload and the sensitization of actin-activated myosin ATPase activity to Ca^{2+} may predispose contractile myocardial cells to the premature contractions.

8.5 Contraction and dilation of smooth muscle

Like cardiac muscle, ACh secreted from postganglionic parasympathetic nerve terminal binds to the muscarinic cholinergic receptors in the plasma membrane of smooth muscle fiber. Muscarinic M_3 receptors are abundant in smooth muscle and glands. Since the M_3 receptor is G_q protein-coupled receptor and mediates an increase in intracellular calcium, it mainly elicits contraction of smooth muscle, such as bronchoconstriction and increases in motility and the tone of stomach and intestine. However, activation of M_3 on vascular endothelial cells induces synthesis of nitric oxide, which diffuses to adjacent vascular smooth muscle cells and causes vasodilatation. The dissociated α subunit of G_q protein is responsible for the stimulation of phospholipase C activity, which catalyzes hydrolysis of one of the membrane phospholipids, phosphatidylinositol 4,5 bisphosphate. The hydrolysis produces one of the major second messenger, inositol triphosphate (IP_3), that is soluble and diffuses inside the cell, where it binds to its receptor, which is IP_3-gated Ca^{2+} channel [6] located in the endoplasmic reticulum (ER). When IP_3 binds its receptor, it increases the release of Ca^{2+} from intracellular stores in the ER into the cytosol, thereby activating the calcium regulated intracellular process. First, the second messenger Ca^{2+} binds to calcium-binding protein calmodulin (CaM) in the cytosol, and then the Ca^{2+}-CaM activates calmodulin-dependent myosin light-chain kinase. This enzyme catalyzes the phosphorylation of the myosin light chain at serine, which induces activation of ATPase in myosin on the actin binding site. Cross-bridges form between myosin and actin, resulting in the contraction of smooth muscle. Constriction of bronchial smooth muscle and increases in motility and the tone of stomach and intestine are especially important among parasympathetic effects on smooth muscles. The bronchial spasm is due to the activation of parasympathetic nervous system. *However, smooth muscles of blood vessels have no parasympathetic innervations except the vessels in skeletal muscle and salivary glands*, and the stimulation of parasympathetic nerve causes vasodilatation in skeletal muscle and salivary glands [7]. Probably diffusion of the nitric oxide produced in adjacent vascular smooth muscle cells overcomes an increase in intracellular calcium.

Norepinephrine (NE) is a neurotransmitter secreted from postganglionic sympathetic nerve terminal, and it binds to the receptors on metabolic G protein-coupled receptors ($\alpha_1, \alpha_2, \beta_1, \beta_2$). NE is a poor stimulant of β_2 adrenergic receptors but is a potent stimulant of α_1, α_2, and β_1 adrenergic receptors. A circulating hormone epinephrine (EPI) is a potent stimulant to α_1, α_2, β_1, and β_2 receptors.

All smooth muscles of blood vessels have innervations of sympathetic nerve and have α_1 adrenergic receptors that induce vasoconstriction. The sympathetic stimulation to α_1 receptor (G_q protein-coupled) causes vasoconstriction resulting from the phosphatidylinositol signal pathway. The effector of dissociated α subunit of G_q protein is the inositol triphosphate-producing enzyme, phospholipase C.

Tracheal and bronchial smooth muscles, gastrointestinal smooth muscles, and blood vessels distributed in the heart, skeletal muscle, lungs, and abdominal viscera have β_2 receptors. The β_2 receptor is G_s protein-coupled receptor. Sympathetic stimulation causes a decrease in the motility and the tone of stomach and intestine resulting from the signal transduction pathway via G_α, that is, the activation of β_2 receptors, adenyl cyclase, and protein kinase A (PKA), successively. Also the stimulation of β_2 receptor in blood vessels causes muscle relaxation called vasodilatation by the same signal transduction pathway via G_α. The activation of PKA results in the phosphorylation of myosin light-chain kinase, which decreases affinity between the light-chain kinase and Ca^{2+}-calmodulin, and the resultant smooth muscle relaxation. Blood vessels of skin and mucosa constrict responding stimulation of sympathetic nerves but show no response to β_2 adrenergic receptor agonists, which indicates that blood vessels of skin and mucosa have only α_1 receptors, probably contributing to maintaining constant body temparature.

The stimulation of sympathetic nerve make tracheal and bronchial smooth muscles dilate probably through the stimulation of EPI secretion from adrenal medulla since human bronchial smooth muscle receives little or no sympathetic innervation, but it contains large number of β_2 adrenergic receptors. EPI is 10–50 times more potent than NE to the β_2 receptors. EPI plays an important role in the relaxation of the bronchial smooth muscle.

After the secretion of ACh, or the administration of muscarinic agonists such as methacholine, carbachol, and pilocarpine, or mechanical stretch, we see increases in the cytosolic Ca^{2+} in smooth muscles. The hydrolysis of one of the membrane phospholipids, phosphatidylinositol 4,5 bisphosphate produces the second messenger, *inositol triphosphate* (IP_3) that increases the oscillatory release of Ca^{2+} from intracellular stores in the endoplasmic reticulum by opening IP_3-gated Ca^{2+} channels, and *diacylglycerol* (DG) that activates protein kinase C (PKC) in the presence of Ca^{2+}. Then the activated PKC subsequently opens L-type Ca^{2+} channels (voltage-gated Ca^{2+} channels) by phosphorylation. ACh raised "the base membrane potential" that probably results from an increase in cytosolic Ca^{2+}. By the way, voltage-gated Ca^{2+} channels in the membrane can be also opened in response to the membrane depolarization brought mechanically by the stretching of the smooth muscle cell [8].

The messenger Ca^{2+} binds to the acidic calcium-binding protein, calmodulin (CaM), and then the activated CaM diffuses and activates calmodulin-dependent myosin light-chain (MLC) kinases. This activated form of the enzyme catalyzes the phosphorylation of the myosin light chain at serine, which induces activation of the ATPase on the actin binding site of the myosin head. The level of actin-activated myosin ATPase activity depends on the degree of myosin light-chain phosphorylation. Both the ATPase activity and the Ca^{2+} sensitivity of the ATPase activity were more remarkable when actin that contained tropomyosin was used to activate the ATPase activity [9]. The data suggests that the actin-activated ATP hydrolysis is enhanced not only by the calcium effects on the phosphorylation of the myosin light chain but also by tropomyosin conferring the Ca^{2+} sensitivity of the actin-activated myosin ATPase. High-energy phosphate released during hydrolysis of ATP catalyzed by the activated ATPase enables the recocking of the myosin head to ready for binding to actin and form the myosin cross-bridge with actin again. The formation of the cross-bridges between myosin and actin is reversible and make muscle contraction repeatable.

On the other hand, the sympathetic stimulation of β_2 receptors causes a decrease in the motility and the muscle tone resulted from activation of adenylyl cyclase, an increase in cyclic AMP, and an activation of protein kinase A (PKA), successively. The smooth muscle relaxation occurs as the activation of PKA results in phosphorylation of myosin light-chain kinase, which decreases affinity between the light-chain kinase and Ca^{2+}-calmodulin. The phosphorylated kinase cannot catalyze the phosphorylation of the myosin light chain, thereby lowering ATPase activity. The process of the relaxation requires a low intracellular Ca^{2+} concentration or low myosin ATPase activity. On the contrary, a high intracellular Ca^{2+} concentration or an increase in myosin light-chain phosphatase activity may restore the muscle tone again.

8.6 Oscillatory contractions of intestine

The muscles of small intestine like other unitary smooth muscles have many gap junctional connections between individual cells and act as one sheet in a coordinated fashion. Thus, the muscle fibers do not need to have an individual innervation, and a depolarization of one fiber triggers synchronous depolarization throughout the muscle and cause continuous partial contractions called the muscle tone. The outstanding characteristic of the small intestinal muscle is its rhythmical segmental and peristaltic contractions. Alternate circular muscle contractions and relaxations at a regular frequency and at regularly spaced intervals are called segmentation movements [10]. Interstitial cells of Cajal (ICC) connected with each other to make cellular network (the submuscular plexus) were identified to be pacemakers [11,12], which may be responsible for generating slow waves of the membrane potential and the rhythmical muscle tone. The rhythmic IP_3-induced calcium release from intracellular stores in the endoplasmic reticulum (ER) was thought to be involved in the pacemaking mechanism in the interstitial cells since blockers of IP_3-induced calcium release abolished the slow-wave activity in the mouse small intestine [13]. Nerve impulses or neurotropic drugs such as ACh and EPI can modulate the activity. A sharp spike superimposed on the small sinusoidal wave activity occurs when the circular muscle contracts. The peristaltic contraction makes rhythmic contractions of the circular muscle without any interruption from the rhythm of segmental contractions.

Each sharp spike may correspond to Na^+ influx, which is necessary for the pulse conduction along the membrane, and the small sinusoidal wave activity may relate to the oscillatory release of stored Ca^{2+} from the ER and be responsible for the rhythmic muscle tone which can be modulated by autonomic nerves, but there is no indication of influx of extracellular Ca^{2+} in a resting state. ACh raised not only "the base membrane potential" probably resulting from an increase in cytosolic Ca^{2+} but also "the sharp spike frequency" suggesting frequent Na^+ influx, which results in large increases in the tonic tension and the rate of the circular muscle contractions as shown in Fig. 8.2. Diacylglycerol (DG), one of hydrolyzed products of phosphatidylinositol 4,5 bisphosphate, activates protein kinase C (PKC) in the presence of Ca^{2+}. The raised base membrane potential probably due to high cytosolic Ca^{2+} is considered as a result of opening of L-type Ca^{2+} channels by the activated PKC. In a relaxation of the smooth muscle, IP_3-gated Ca^{2+} channels are not open but in a leak state, and voltage-gated Ca^{2+} channels in the plasma membrane are thought to be in a closing state, resulting in low base membrane potential and reduced sinusoidal wave activity, and the myosin ATPase is inactivated, reducing the frequency of rhythmic muscle contractions. The relaxation of the smooth muscle is seen when sympathetic activity dominates parasympathetic activity.

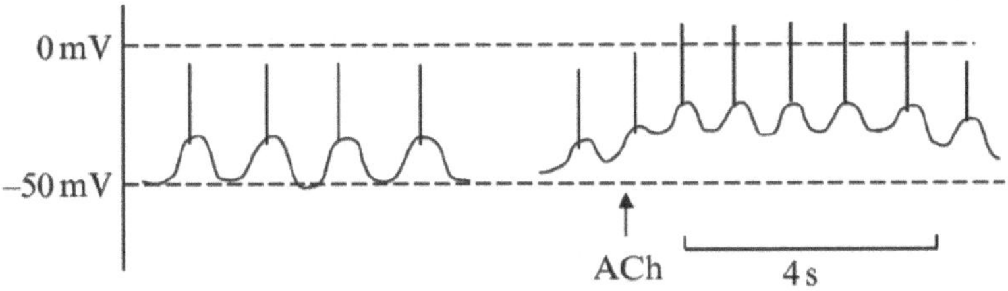

Fig. 8.2 Effects of ACh on the membrane potential of small intestinal smooth muscle in vitro. The sharp spike potential occurs when the circular muscle contracts, and it is superimposed on the slow-wave activity which may relate to the oscillatory release of stored Ca^{2+} from the ER and be responsible for the rhythmic muscle tone. Since the slow waves are small, they represent no indication of influx of extracellular Ca^{2+}. ACh raised not only "the base membrane potential" probably resulting from an increase in cytosolic Ca^{2+} due to extracellular Ca^{2+} influx but also "the sharp spike frequency" suggesting frequency of Na^{+} influxes, which results in large increases in the tonic tension and the rate of the circular muscle contractions.

Coordinated movements of the stomach and intestines are required to digest and propel intestinal contents along the digestive tract. The complex patterns of contraction and relaxation necessary for proper motility of the gastrointestinal (GI) tract are controlled by the nerves within the GI walls. Random, unorganized motor activity with peristaltic and antiperistaltic complexes occasionally occurs during feeding, and this allows gastric remixing of foods. After this, more regular activity begins with the peristaltic waves that allows progression of undigested food through the intestines again.

8.7 Vasodilation

The widening of the vascular bed decreases resistance in segments of comparable length. Resistance vessels are small arteries and arterioles chiefly, and capillaries and smallest venules. In most peripheral part of the circulation, flow is controlled so that with elevation of tissue metabolism flow increases, i.e., the vessels of digestive tracks dilate after eating, and the vessels of brain dilate during study. Local cardiovascular control mechanisms control the radius of peripheral arteries and influence cardiac output. They have been shown to operate in the absence of central nervous system and peripheral autonomic innervation. The smallest blood vessels have an inherent automaticity, which can be modulated by the stimulation of nerve fibers, circulating hormone such as epinephrin, or local electric stimulation. The mechanism of the inherent automaticity probably is due to the rhythmic

Ca^{2+} release from the ER, relating to the basic muscle tone. All smooth muscles of blood vessels have innervations of sympathetic nerve which modulate the muscle tone. Stimulation of β_2 receptor causes vasodilatation. The administration of β_2 adrenergic receptor agonists or α adrenergic receptor antagonists relaxes small blood vessels.

There is another cellular mechanism that leads to vasodilatation. Activation of M_3 on vascular endothelial cells induces synthesis of nitric oxide which diffuses to adjacent vascular smooth muscle fibers and causes vasodilatation. Endothelial-derived relaxing factor was later identified as the gaseous second messenger, nitric oxide (NO). Nitric oxide molecules are free radicals that diffuse freely across the cell membranes and induce the smooth muscle of blood vessels to relax by activating guanylate cyclase, which catalyzes the formation of highly reactive second intracellular messenger of cyclic GMP from GTP (guanosine triphosphate). Cyclic GMP can activate cyclic GMP-dependent protein kinases [14]. Cyclic GMP-dependent protein kinases stimulate the opening of Ca^{2+}-activated K^+ channels [15], leading to hyperpolarization and relaxation of the muscle fibers in a resting state, and blocks the activity of phospholipase C which catalyzes the formation of inositol triphosphate (IP_3) from membrane phospholipid, phosphatidylinositol 4,5 diphosphate (PIP_2) by hydrolysis. The decreased amount of hydrolyzed product, IP_3 reduces the discharge of the stored calcium ions from the endoplasmic reticulum, leading to vasodilatation.

This signaling pathway is naturally lowered by the action of phosphodiesterase which catalizes second messenger, cyclic GMP into a nonsignaling form, GMP. The drugs such as sidenafil and vardenafil are all specific inhibitors of phosphodiesterase type-5, which is found mainly in the smooth muscle cells lining the blood vessels supplying the corpus cavernosum of the penis. Oral administration of these drugs blocks the action of phosphodiesterase type-5, thereby increasing the blood flow in the limited region and offsetting erectile dysfunction.

8.8 Bronchoconstriction and bronchodilation

In healthy lungs, ACh produced and released by both neuronal and non-neuronal sources acts on muscarinic receptors to regulate physiological functions. Muscarinic M_3 receptors control smooth muscle tone, mucus secretion, and inflammation. In chronic obstructive disease and asthma, cholinergic mechanisms contribute to increased bronchoconstriction and mucus secretion that limit airflow [16].

Human bronchial smooth muscle receives little or no sympathetic innervation, but it contains large number of β_2 adrenergic receptors [17]. The stimulation of β_2 adrenergic receptors (Gs protein-coupled) leads to the activation of adenylyl cyclase, increased cyclic AMP, and the activation of cyclic AMP-dependent protein kinase (PKA), which results in phosphorylation of myosin light-chain kinase, hence, decreased affinity of the light-chain kinase to the Ca^{2+}-calmodulin, resulting in smooth muscle relaxation [18], which can be also seen in blood vessels. β_2 adrenergic receptor agonists also increase Ca^{2+}-activated K^+ currents in airway smooth muscle, leading to membrane hyperpolarization and relaxation in a resting state. Since epinephrine (EPI) is 10–50 times more potent than norepinephrine (NE) on the β_2 receptors, circulating hormone, EPI may play an important role in the relaxation of bronchial smooth muscle during sympathetic stimulation.

Short-acting β_2 adrenergic agonists such as albuterol (salbutamol) are used to increase airflow in conditions such as asthma and chronic obstructive pulmonary disease (COPD). Muscarinic antagonist medications provide additional benefit when used in combination with albuterol in patients with moderate or severe symptoms. Muscarinic antagonists, i.e., ipratropium bromide increase airflow in asthma by blocking cholinergic tone and by blocking reflex bronchoconstriction mediated by the vagus nerves. They may also inhibit secretion and enhance clearance of mucus. In asthma, airflow limitation is generally fully reversible and is caused by bronchoconstriction. However, in case of COPD, the airflow limitation is not fully reversible, and is caused by structural changes, i.e., narrowing of peripheral airways along with the parenchymal destruction. Muscarinic antagonists can increase airflow in COPD by blocking cholinergic tone at the airway smooth muscle, but effects are limited.

8.9 Conclusions

The single unit smooth muscles and cardiac muscle do not have individual innervations but have gap junctional connections between individual cells and acts as one sheet in a coordinated fashion in contrast to skeletal and the multiunit smooth muscles. The depolarization in the pacemaker cells initiates synchronous depolarizations, which can be modulated by the autonomic nervous system. In the heart, sympathetic stimulation can raise cAMP and Ca^{2+} levels in myocardial cells and increase excitability of the cells by activating the actin-activated ATPase on the myosin head. If ATPase in the myosin head becomes more active at lower concentration of Ca^{2+}, the muscle fibers would get more excitable and induce arrhythmias. On the other hand, ACh from the vagus nerve to pacemaker cells decreases cAMP and Ca^{2+} levels and causes hyperpolarization in a resting state and reduces the pulse of heart through signal transduction pathways via G_{α} and $G_{\beta\gamma}$.

Since the M_3 receptor is G_q-coupled and mediates an increase in intracellular calcium, it mainly elicits contraction of smooth muscles such as increases in motility and the tone of stomach and intestine, and bronchoconstriction. However, activation of M_3 on vascular endothelial cells induces synthesis of nitric oxide which diffuses to adjacent smooth muscles of blood vessels and overcomes the increase in intracellular calcium and causes vasodilatation. The outstanding characteristic of the small intestinal muscle is its rhythmicity. The basic source of pacemaker activity is thought to be rhythmic Ca^{2+} diffusion through IP_3-gated Ca^{2+} channels from the ER, which produces tone of the smooth muscle. However, the pacemaker activity can be modified by parasympathetic stimulation which causes an increase in cytosolic Ca^{2+} via opening of IP_3-gated Ca^{2+} channels and voltage-gated Ca^{2+} channels, thereby activating myosin ATPase and by sympathetic stimulation which causes a decrease in affinity between the light-chain kinase and Ca^{2+}-calmodulin, thereby inactivating myosin ATPase. The second messenger in this case is triphosphate, diacylglycerol and Ca^{2+} instead of cyclic AMP and Ca^{2+} in the heart. The increase in cytosolic Ca^{2+} leads to the phosphorylation of the myosin light chain at serine, which induces the activation of myosin ATPase in the actin binding site.

Muscarinic M_3 receptors control typically smooth muscle tone, mucus secretion, and inflammation. Human bronchial smooth muscle receives little or no sympathetic innervation, but it contains large number of β_2 adrenergic receptors. In healthy lungs, ACh by both neuronal and nonneuronal sources acts through muscarinic receptors to regulate physiological functions. Muscarinic antagonist such as ipratropium bromide medications provide additional benefit when used in combination with short-acting β_2 adrenergic agonists such as albuterol in patients with moderate or severe symptoms. In chronic obstructive pulmonary disease (COPD) and asthma, cholinergic mechanisms contribute to increased bronchoconstriction and mucus secretion that limit airflow. Muscarinic antagonists such as ipratropium bromide increase airflow in asthma by blocking cholinergic tone and by blocking reflex bronchoconstriction mediated by the vagus nerves. They may also inhibit secretion and enhance clearance of mucus. However, in case of COPD, airflow limitation is not fully reversible and is caused by structural changes, i.e., narrowing of peripheral airways along with parenchymal destruction. Muscarinic antagonists can increase airflow in COPD by blocking cholinergic tone at airway smooth muscle, but the effects are limited.

References

[1] K. Barrett, S. Barman, S. Boitano, H. Brooks, Excitable Tissue: Muscle, Review of Medical Physiology, twenty fourth ed., McGraw-Hill, New York, 2012 (Chapter 5).
[2] M.D. Bootman, Calcium signaling, Cold Spring Harb. Perspect. Biol. (2012), https://doi.org/10.1101/Cshperspect.a011171.
[3] R. Zucchi, S. Ronca-Testoni, The sarcoplasmic reticulum Ca2+ channel/ryanodine receptor: modulation by endogenous effectors, drugs and disease states, Pharmacol. Rev. 49 (1) (1997) 1–52.
[4] J.M. Chalovich, E. Eisenberg, Inhibition of actomyosin ATPase activity by troponin-tropomyosin without blocking the binding of myosin to actin, J. Biol. Chem. 257 (5) (1982) 2432–2437.
[5] D. Szczesna, R. Zhang, J. Zhao, M. Jones, G. Guzman, J.D. Potter, Altered regulation of cardiac muscle contraction by troponin T mutations that cause familial hypertrophic cardiomyopathy, J. Biol. Chem. 275 (1) (2000) 624–630.
[6] T.C. Westfall, D.P. Westfall, Neurotransmission: The Automatic and Somatic Motor Nervous Systems, the Pharmacological Basis of Therapeutics, eleventh ed., McGraw-Hill, New York, 2006 (Chapter 6).
[7] T.C. Westfall, D.P. Westfall, Drugs acting at synaptic and neuroeffector junctional sites, in: The Pharmacological Basis of Therapeutics, eleventh ed., McGraw-Hill, New York, 2006, pp. 137–181.
[8] R.C. Webb, Smooth muscle contraction and relaxation, Adv. Physiol. Educ. 27 (2003) 201–206.
[9] S. Chacko, A. Rosenfeld, Regulation of actin-activated ATP hydrolysis by arterial myosin, Proc. Natl. Acad. Sci. 79 (2) (1982) 292–296.
[10] N.C. Hightower, H.D. Janowitz, Movements of the alimentary canal, in: Best & Taylor's Physiological Bases of Medical Practice, ninth ed., The Williams & Wilkins, Baltimore, MD, 1967, pp. 111–121.
[11] S.D. Koh, K.M. Sanders, S.M. Ward, Spontaneous electrical rhythmicity in cultured interstitial cells of Cajal from the murine small intestine, J. Physiol. 513 (1998) 203–213.
[12] L. Thomsen, T.L. Robinson, J.C. Lee, L.A. Farraway, M.J. Hughes, D.W. Andrews, J.-D. Huizinga, Interstitial cells of Cajal generate a rhythmic pacemaker current, Nat. Med. 4 (1998) 848–851.
[13] J. Malysz, G. Donnelly, J.D. Huizinga, Regulation of slow wave frequency by IP(3)-sensitive calcium release in the murine small intestine, Am. J. Physiol. Gastrointest. Liver Physiol. 280 (2001) G439–G448.
[14] L.J. Ignarro, Nitric Oxide: A Unique Endogenous Signaling Molecule in Vascular Biology, 1998, Nobel Lecture, Available from: www.nobelprize.org/nobel_prizes/medicine/laureates.
[15] B.D. Kyle, et al., Specific phosphorylation sites underlie the stimulation of a large conductance, Ca^{2+}-activated K^+ channel by cGMP-dependent protein kinase, FASEB J. 27 (5) (2013) 2027–2038.
[16] K.S. Buels, A.D. Fryer, Muscarinic receptor antagonists: effects on pulmonary function, Handb. Exp. Pharmacol. 208 (2012) 317–341.
[17] B.J. Undem, Pharmacotherapy of asthma, in: The Pharmacological Basis of Therapeutics, eleventh ed., McGraw-Hill, New York, 2006, pp. 720–721.
[18] J.R. Miller, P.J. Silver, J.T. Stull, The role of myosin light chain kinase phosphorylation in beta-adrenergic relaxation of tracheal smooth muscle, Mol. Pharmacol. 24 (2) (1983) 235–242.

CHAPTER 9

Gas exchange and respiratory insufficiency

9.1 Introduction

Oxygen diffuses rapidly from the alveoli into blood in the pulmonary capillaries, and carbon dioxide diffuses in the opposite direction. Since oxygen concentration is always higher in the blood capillaries than in the tissue cells, it diffuses continuously from blood to the tissue cells via the interstitial fluid in one direction. The concentration of carbon dioxide is higher in the tissue cells, so it diffuses in the opposite direction of oxygen. Oxygen is carried in the high-speed blood flow in the vessel in two ways, i.e., (1) loosely bound to hemoglobin within red blood cells and (2) dissolved in the plasma. Most of the oxygen is carried in the form of oxyhemoglobin. At high altitudes, the oxygen concentration in the blood plasma decreases, but the total concentration of oxygen in the blood is almost constant since hemoglobin in blood cells is almost saturated whenever the plasma partial pressure of oxygen is above 70 mmHg. Dalton's law of partial pressures and Henry's law will help understand the movements of gases into and out of the blood.

Dalton's law states that at any specified temperature total pressure exerted by a gas mixture is the sum of pressures exerted independently by each constituent gas. The pressure exerted by each constituent called partial pressure is directly proportional to the percentage of the constituent in the mixture as described below.

Consider a mixture of three gases described by mole numbers n_1, n_2, n_3 in the container of volume V at temperature T.

$$n_{\text{total}} = n_1 + n_2 + n_3$$

$$p = \frac{n_{\text{total}} R T}{V} \tag{9.1}$$

where p is total pressure.

We define the partial pressure of each constituent as the pressure that the constituent would exert as if it were alone in the container of volume V at

https://doi.org/10.1016/B978-0-323-98803-2.00009-2

temperature T. The partial pressures of each constituent gas are denoted by p_1, p_2, p_3, and the total pressure is p.

The partial pressure of each constituent is defined as follows.

$$p_1 = n_1\frac{RT}{V},\quad p_2 = n_2\frac{RT}{V},\quad p_3 = n_3\frac{RT}{V} \tag{9.2}$$

$$p = n_{\text{total}}\frac{RT}{V} = (n_1 + n_2 + n_3)\frac{RT}{V} = p_1 + p_2 + p_3 \tag{9.3}$$

Eq. (9.3) shows that the total pressure is the sum of the partial pressure of each constituent.

From Eqs. (9.1) and (9.2),

$$\frac{p_1}{p} = \frac{n_1 RT}{V}\frac{V}{n_{\text{total}}RT} = \frac{n_1}{n_{\text{total}}} = X_1 \tag{9.4}$$

$$p_1 = X_1 p,\quad p_2 = X_2 p,\quad p_3 = X_3 p \tag{9.5}$$

Eq. (9.5) shows that the partial pressure of each constituent of the gas mixture is the product of mole fraction, X_i and total pressure.

The partial volume of each constituent is defined as follows:

$$V_i = n_i\frac{RT}{p}$$

$$V = n_{\text{total}}\frac{RT}{p} = (n_1 + n_2 + n_3)\frac{RT}{p} = V_1 + V_2 + V_3$$

$$X_i = \frac{n_i}{n_{\text{total}}} = \frac{pV_i}{RT}\cdot\frac{RT}{pV} = \frac{V_i}{V} \tag{9.6}$$

where V is the total volume of the gas mixture, V_i is the partial volume of a constituent gas in the gas mixture. The mole fraction, X_i is equal to the volume fraction in the mixture (see Eq. 9.6).

The partial pressure can be calculated by measuring the volume fraction in a sample since we know that the mole fraction is equal to the volume fraction in the mixture.

At sea level, atmospheric pressure is 760 mmHg.

In the inhaled air, the volume fraction of oxygen and carbon dioxide are 0.209 and 0.31, respectively.

$$p_{O_2} = \frac{V_i}{V}p = 0.209 \times 760\,\text{mmHg} = 159\,\text{mmHg}$$

$$p_{CO_2} = \frac{V_i}{V}p = 0.0004 \times 760\,\text{mmHg} = 0.31\,\text{mmHg}$$

The air in the alveoli contains less oxygen, more carbon dioxide and water vapor. The volume fraction of oxygen and carbon dioxide are 0.137 and 0.52, respectively, at end-expiratory and end-inspiratory positions.

$$p_{O_2} = \frac{V_i}{V} p = 0.137 \times 760\,\text{mmHg} = 104\,\text{mmHg}$$

$$p_{CO_2} = \frac{V_i}{V} p = 0.052 \times 760\,\text{mmHg} = 39.5\,\text{mmHg}$$

Henry's law says that the amount of dissolved gas in a liquid is proportional to its partial pressure above the liquid. The proportionality factor is called Henry's law constant. The partial pressure of a solute in the vapor phase above the liquid is proportional to its mole fraction in the liquid in equilibrium. The partial pressure of a gas increases when its volume fraction in the vapor phase or total pressure is increased. The greater the partial pressure of a gas, the more and the faster the gas will dissolve in the liquid [1].

$$p_i = \frac{V_i}{V} p = K_i \frac{n_i}{n_{\text{liquid}}} \tag{9.7}$$

where p_i is the partial pressure of dissolved gas in the vapor phase and $\frac{V_i}{V}$ is the volume fraction of the gas. K_i is Henry's law constant. Each different gas has individual Henry's constant. The constant of nitrogen is 2 times as much as that of oxygen, and the constant of carbon dioxide is 1/20 times as much as that of oxygen, respectively. Carbon dioxide is soluble in the plasma 20 times as much as oxygen in the plasma when the two gases have the same partial pressure in the vapor phase. The n_i is the mole number of dissolved gas, n_{liquid} is the total mole number of the solution. The partial pressure of a gaseous molecule in the vapor phase is proportional to the amount of the gas dissolved in the liquid such as the blood plasma and lymph. The partial pressure of oxygen and carbon dioxide increase when water pressure is large under the sea and decrease when atmospheric pressure is low at a high altitude, and the amount of the gas dissolved in the plasma is proportional to them.

Hemoglobin (Hb) is an oxygen transport protein, and there are millions of hemoglobin molecules in a red blood cell. Each hemoglobin molecule can combine with four molecules of oxygen. The oxygen bound hemoglobin called oxyhemoglobin is written as HbO_2. Hemoglobin that has released oxygen is called reduced hemoglobin and is denoted by HHb. Loading and unloading of oxygen are reversible.

$$HHb\,(aq) + O_2\,(aq) \leftrightarrow HbO_2\,(aq) + H^+$$

At a high altitude of 3 km where the atmospheric pressure is 525 mmHg, the amount of O_2 dissolved in the plasma of arterial blood might drop to 69% of that at sea level according to Henry's law as the partial pressure of oxygen in the alveoli drops 69% of that at sea level. How can the body compensate for the people living at high altitude? The hemoglobin in red blood cells remains almost saturated with oxygen at this level of O_2 in blood plasma. The oxygen delivery to the tissues is still adequate. However, the body produces more hemoglobin for the people living at a high altitude to adapt them to any situation in the atmosphere.

When we are exposed to carbon monoxide, it binds to hemoglobin in red blood cells replacing oxygen. Inhalation of pure oxygen at 1.5 atm is used for the treatment of patients suffering from carbon monoxide poisoning to expel carbon monoxide binding to hemoglobin by means of dissolving a large amount of oxygen in the blood plasma.

Under the deep sea, more nitrogen dissolves in the blood plasma of a scuba diver due to the high partial pressure of nitrogen in the alveoli since the total pressure of the vapor in the alveoli is high due to the high-water pressure. The excessive nitrogen in the plasma and tissue fluid forms bubbles and causes decompression sickness if the diver comes out rapidly of the deep depth.

However, the solubility of a gas in the liquid depends not only on its partial pressure but also the kind of gas dissolved and temperature. Carbon dioxide is the most soluble atmospheric gas in water. Carbon dioxide dissolves 20 times as much as oxygen in the plasma when the gases have the same partial pressure in the vapor phase. Nitrogen is dissolved only half as much as oxygen. Increasing temperature decreases gas solubility.

9.2 Gas exchange between blood and organs

It is important to make sure that p_{O_2} in blood or tissues indicates the amount of oxygen dissolved in the plasma or tissue fluid. The alveoli p_{O_2} and the pulmonary capillary p_{O_2} is not much different at equilibrium. The same is true for p_{CO_2}. Normal values for alveolar, arterial, and venous gas pressures are presented in Table 9.1. The alveolar values have been calculated from arterial values, assuming alveolar-arterial pressure differences (−2 for CO_2, +13 mmHg for O_2). The CO_2 and O_2 pressures in venous and arterial

Table 9.1 Volume percentages and partial pressures of gases in the atmosphere and the alveoli, artery, and vein at the end-expiratory position.

	Air at sea		Alveoli		Artery		Vein	
Gas	**V%**	**mmHg**	**V%**	**mmHg**	**V%**	**mmHg**	**V%**	**mmHg**
CO_2	0.04	0.3	5.2	39.5	5.57	42	6.50	46
O_2	20.9	159	13.7	104	12.1	91	5.80	41
N_2	78.6	597	74.9	569	76.1	574	81.1	574
H_2O	0.46	3.5	6.20	47.1	6.23	47	6.64	47
Total	100	759.8	100	759.6	100	754	100	708

blood are based on direct measurements. For N_2 and H_2O, the pressures in the alveoli, artery, and vein are assumed to be the same.

Arterial p_{O_2} lower than 55 mmHg is considered as the insufficient oxygenation of blood. Arterial p_{CO_2} higher than 45 mmHg is related to hypoventilation. Arterial p_{CO_2} lower than 35 mmHg is related to hyperventilation.

9.2.1 Gas exchange between blood and lungs

Partial pressure gradients of oxygen and carbon dioxide between the blood in the pulmonary capillaries and alveoli in the lungs cause diffusion of gases across the respiratory membrane. In the alveoli of the lungs, p_{O2} is 104 mmHg, and p_{CO2} is 40 mmHg, whereas the plasma p_{O2} is 41 mmHg, and p_{CO2} is 46 mmHg in the blood leaving tissues and entering the lungs. As a result, oxygen diffuses rapidly from the alveoli into the venous blood in the pulmonary capillary, and carbon dioxide diffuses in the opposite direction.

9.2.2 Oxygen saturation of Hemoglobin

Oxygen is carried in blood in two ways, i.e., being loosely bound to hemoglobin within red blood cells or dissolved in plasma. Since oxygen is poorly soluble in water, only about 1.5% of the oxygen is carried in the dissolved form. Every 100 mL of systemic arterial blood contains about 20 mL of oxygen. Its concentration is written as 20 vol%. However, the plasma contains only 0.3 vol% of oxygen. That means a majority of oxygen is carried by red blood cells from the lungs to the tissues in the form of oxyhemoglobin. The average p_{O_2} of the blood plasma leaving lungs is 91 mmHg which comes from 13.7 V% (104 mmHg) of oxygen in the alveoli at the end-expiratory and end-inspiratory positions, plus alveolar-arterial pressure differences of 13 mmHg. The arterial p_{O_2} does not indicate the O_2 saturation of hemoglobin which relates to hypoxia, since p_{O_2} in the blood indicates the amount of oxygen dissolved in the plasma, but not in whole blood. However, values of arterial p_{O_2} in the plasma are a good index of pulmonary function. Significantly low arterial p_{O_2}, lower than 55 mmHg, is considered to indicate the insufficient oxygenation of the blood and suggests pulmonary insufficiency seen in pneumonia.

Hemoglobin is almost saturated when plasma p_{O_2} is over 70 mmHg. Every 100 mL of systemic arterial blood contains about 20 mL of oxygen. Resting tissues consume about 25% of Hb-bound oxygen which is about

5 mL of O_2 per 100 mL of blood. And then plasma p_{O_2} drops from 91 mmHg to 41 mmHg, but blood leaving tissues contains still 15 vol% of oxygen (15 mL of O_2 per 100 mL of blood) since oxygen saturation of Hb drops only 25% (Fig. 9.1). On the other hand, a small drop of the plasma p_{O_2} below 50 mmHg results in a large increase in O_2 unloading. Exciting tissues consume about 70% of Hb-bound oxygen which is about 14 mL of O_2 per 100 mL of blood. And plasma p_{O_2} drops from 91 mmHg to 20 mmHg, and blood leaving tissue contains only 6 mL of O_2 per 100 mL of blood.

High temperature, low plasma pH, and high plasma p_{CO_2} decrease hemoglobin saturation at a certain plasma p_{O_2}. As cells metabolize glucose and consume O_2, they release CO_2 which increases the tissue p_{CO_2}, H^+, and heat. As a result, more oxygen is unloaded in the hard-working tissues [2].

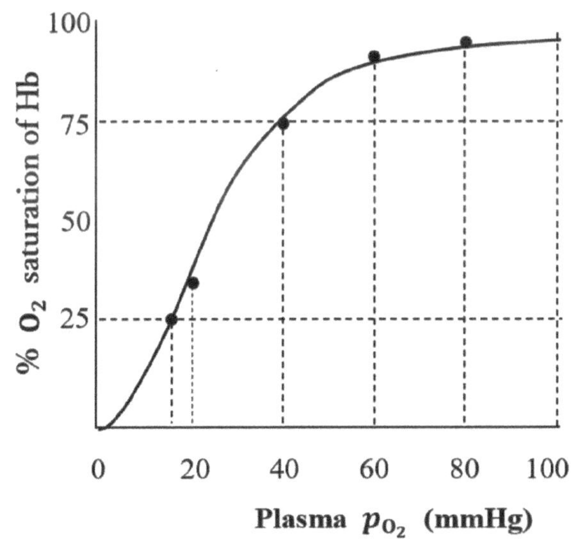

Fig. 9.1 Oxygen-hemoglobin dissociation curve. Hemoglobin is almost saturated when plasma p_{O_2} is above 70 mmHg. Every 100 mL of systemic arterial blood contains about 20 mL of oxygen. Resting tissues consume about 25% of Hb-bound oxygen which is about 5 mL of O_2 per 100 mL of blood. And then plasma p_{O_2} drops from 91 mmHg to 41 mmHg, but blood leaving tissues contains still 15 vol% of oxygen (15 mL of O_2 per 100 mL of blood) since oxygen saturation of Hb drops only 25%. A small drop of plasma p_{O_2} below 40 mmHg results in a large increase in O_2 unloading. Exciting tissues consume about 70% of Hb-bound oxygen which is about 14 mL of O_2 per 100 mL of blood. And plasma p_{O_2} drops from 91 mmHg to 20 mmHg, and blood leaving tissue contains only 6 mL of O_2 per 100 mL of blood.

9.2.3 Gas exchange between blood and tissues

Tissue fluid p_{O2} is slightly larger than 40 mmHg and p_{CO2} is slightly larger than 46 mmHg, whereas in blood leaving lungs and entering tissue capillaries, plasma p_{O2} is 91 mmHg and p_{CO2} is 42 mmHg. As a result, oxygen diffuses rapidly from the capillary blood into tissues, and carbon dioxide diffuses in the opposite direction.

Example 9.1 (1) A male patient, age 65 with the chronic obstructive pulmonary disease did not respond when his family spoke to him and was taken to the hospital by ambulance. The results of the arterial blood gas analysis were p_{O2} 52.2 mmHg, p_{CO2} 72.2 mmHg, and pH 7.26.

Low p_{O2} indicates that he is hypoxia. High p_{CO2} indicates hypoventilation. The pH that is lower than the normal range of 7.35–7.45 indicates acidosis due to excessive dissolved CO_2 in the plasma. (2) A male patient, age 21 was suffering from fatigue, dry mouth, and frequent urination. He was lying unconscious on the floor and was taken to the hospital by ambulance. He was at once diagnosed as diabetic by medical examination. The results of the atrial blood gas analysis were p_{O2} 98 mmHg, p_{CO2} 14.3 mmHg, % saturation of O_2 100%, pH 6.98, and BE −21.2 mEq/L. Base excess (BE) is defined as the amount of strong acid that must be added to each liter of fully oxygenated blood to normalize the pH to 7.40 at a temperature of 37°C and p_{CO2} of 40 mmHg. A negative base excess can be defined in terms of the amount of strong base that must be added to normalize the pH to 7.04. Negative BE corresponds to a deficiency of base, mainly HCO_3^-.

The pH 6.98 and BE −21.2 mEq/L indicate metabolic acidosis. Low p_{CO2} suggests a compensation of acidosis by hyperventilation.

9.3 Regulation of respiration

Respiration is controlled by two separate mechanisms, i.e., the voluntary system located in the cerebral cortex and the other automatic system driven by a group of pacemaker neurons in the medulla [3]. Rhythmical impulses from the pacemaker cells activate motor neurons in the cervical and thoracic spinal cord that innervate inspiratory muscles. Those in the cervical cord activate the diaphragm via the phrenic nerves, and those in the thoracic spinal cord activate the external intercostal muscles. A rise in p_{CO_2} and H^+ level in arterial blood or a remarkable decrease in arterial p_{O_2} increases the level of the respiratory neuronal activity in the medulla via chemoreceptors in the carotid and aortic bodies, leading to an increase in ventilation. Changes in the p_{CO_2} and H^+ level in the opposite direction effect the respiration slightly negatively.

9.4 Diffusion across respiratory membrane

The thin respiratory membrane separates air within the alveoli from the blood within pulmonary capillaries. It consists of the alveolar epithelium, the fused basement membrane between the alveolar epithelium and the capillary endothelium, and the capillary endothelium. Gases must be dissolved in a surface phospholipid film that is surfactant covering the alveolar lining maintaining the lung elasticity, before crossing the alveolar epithelium, the interstitial fluid, and the capillary endothelium to enter bloodstream. Normally, the distance of the path to diffuse is very short, 0.1 μm, but in diseases, it may be longer as the fused basement membranes are separated by the interstitial edema fluid and exudate or may be replaced by fibrous tissue [4]. Differences between partial pressures of oxygen and carbon dioxide in the alveoli of the lungs and those in the blood within the pulmonary capillaries cause diffusion of the gases across the respiratory membrane along their concentration gradients, though the speed of diffusion across the membrane is slow. In the pathological states, the speed of diffusion decreases. Once getting in the capillary, gases dissolved in the plasma spread in the bloodstream where the speed of diffusion is fast.

9.5 Airflow in the lungs and pulmonary resistance

The thoracic cavity that is bounded below by the diaphragm, is enclosed by the sternum, ribs, and thoracic vertebrae, and contains the lungs, the heart, and esophagus. The volume of the thoracic cavity changes according to the rhythmical alternate contractions and relaxations of the diaphragm and the external intercostal muscles (Fig. 9.2), which is controlled by the pacemaker cells in the medulla.

The respiratory passageway in the lung includes the trachea, bronchi, bronchioles, and alveolar ducts. The passage smaller than 1 mm in diameter is called a bronchiole, and the thin-walled air sacs are called alveoli. Constituents of the air in the alveoli are different from those in the external atmosphere because of the partial gas exchange in the lungs as shown in Table 9.1. The air in the alveoli is a mixture of newly inspired air and the air remaining in the respiratory passageways and alveoli between breaths. The inhaled air consists almost entirely of oxygen and nitrogen whereas the air in the alveoli

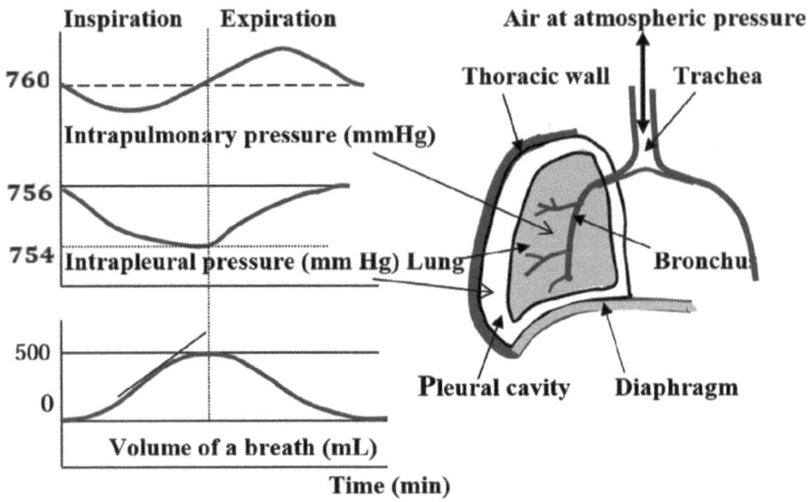

Fig. 9.2 Changes in intrapulmonary and intrapleural pressure during inspiration and expiration. The volume of the thoracic cavity increases during inspiration. As intrapleural pressure decreases, intrapulmonary pressure decreases below the atmospheric pressure at the first half stage of inspiration, but during the last half stage, the pressure inside the lungs increases as the air goes in and returns to the atmospheric pressure at the end-inspiratory position when an inward force of the recoil of lungs balances to the intrapleural pressure, and the airflow into the lungs and the volume change of lungs stop, and then lung volume reaches the maximum. The volume of the thoracic cavity decreases during expiration. As the intrapleural pressure increases, the intrapulmonary pressure increases above the atmospheric pressure at the first half stage of expiration, but during the last half stage, the pressure inside the lungs decreases as the air goes out and returns to the atmospheric pressure at the end-expiratory position when the airflow going out of the lungs and the volume change of lungs stop, and then the lung volume reaches the minimum. The tangent of tidal volume curve (mL/min) indicates the airflow in the trachea. The airflow becomes zero at the end-expiratory and end-inspiratory positions when intrapulmonary pressure is equal to atmospheric pressure.

contains more carbon dioxide and water vapor and much less oxygen. These differences are created because of the partial gas exchange in the lungs. The air in the alveoli is the mixture of the inhaled air and the remaining air in the lungs, which is the sum of the oxygen, the carbon dioxide and the unchanged nitrogen from the pulmonary venous blood, and the humified inhaled air in the air passageways and sacs. Only an average of 500 mL of atmospheric air is inspired with each respiration cycle in adults. The alveolar partial pressure of oxygen and carbon dioxide may change according to hyperventilation or hypoventilation.

Thoracic cavity volume increases when contraction of the dome-shaped diaphragm makes it move inferiorly and flatten out while contraction of the external intercostal muscles lifts the rib cage and pulls the sternum superiorly. During inspiration, a decrease in intrapleural pressure is created by increasing the volume of the thoracic cavity. As intrapleural pressure decreases, intrapulmonary pressure that is the average pressure inside the lungs decreases below the atmospheric pressure at the first half stage of inspiration, but during the last half stage, the pressure inside the lungs increases as the air goes in and returns to the atmospheric pressure at the end-inspiratory position when an inward force of the recoil of lungs balances to the intrapleural pressure, and the airflow into the lungs and the volume change of lungs stop, and then the lung volume reaches the maximum as seen in Fig. 9.2.

During expiration, the volume of the thoracic cavity decreases when the diaphragm and the external intercostal muscles relax. As the intrapleural pressure increases, the intrapulmonary pressure increases above the atmospheric pressure at the first half stage of expiration, but during the last half stage, the pressure inside the lungs decreases as the air goes out and returns to the atmospheric pressure at the end-expiratory position when the airflow going out of the lungs and the volume change of lungs stop, and then the lung volume reaches the minimum as seen in Fig. 9.2.

The gas flow into and out of the alveoli in the lungs is directly proportional to the intrapulmonary pressure that is the average pressure in the alveoli if the atmospheric pressure is constant. The intrapulmonary pressure can be estimated, measuring pressure changes in the body plethysmograph (a box with a sensitive pressure gauge and a patient in the air-tight box) where a patient is placed. If there is no obstruction in the respiratory passageways, the very small intrapleural pressure change of 2 mmHg can produce about 500 mL of gas exchange per breath in adults. In healthy lungs, airway resistance is created by friction between molecules of air and walls of bronchi and bronchioles. It is dangerous to measure intrapleural pressure directly by cannulation because of the possibility of a lung puncture, but a satisfactory approximation of changes in intrapleural pressure can be obtained by measuring pressure in the esophagus through a small balloon with the patient sitting since the esophagus is in the thoracic cavity [5].

The rate of diffusion through airway passages is fast. The magnitude of flow depends on the pressure difference and resistance of the airway since the pressure change indicates the concentration change of air molecules in the airway passage. If pressure difference exists in the airway passage in the lungs, the flow will go in the direction to minimize the pressure difference.

If atmospheric pressure is the same where a patient breathes, the magnitude of the airflow through the airway passage in the lungs is proportional to the pressure change of the intrapulmonary pressure, and inversely proportional to pulmonary resistance in the respiratory airways that include trachea, bronchi, bronchioles, and alveolar ducts. The constriction of tracheal smooth muscle and formation of plugs by such as inflammatory exudate or pus in the respiratory passage makes the airway resistance high, which means that in order to exchange the same amount of the gases, more pressure difference between the atmosphere and the alveoli should be produced by contraction of the diaphragm and the external intercostal muscles through coughs or deep forced breaths.

$$R = \frac{\Delta p}{F} \tag{9.8}$$

where F is airflow through the trachea, Δp is the pressure difference between atmospheric pressure and the average pressure in the alveoli, and R is pulmonary resistance in the airway.

In pneumonia, the partial lung collapse occurs as some air sacs called alveoli fill with inflammatory exudate or pus, which can be observed as the diffuse opacity in chest X-ray, and the ventilation is impaired as oxygen is hard to reach the bloodstream due to a decrease in the diffusion capability in some alveoli (Fig. 9.3). Mechanical ventilation is required to improve the impaired gas exchange. Short-acting β_2 adrenergic agonists such as albuterol

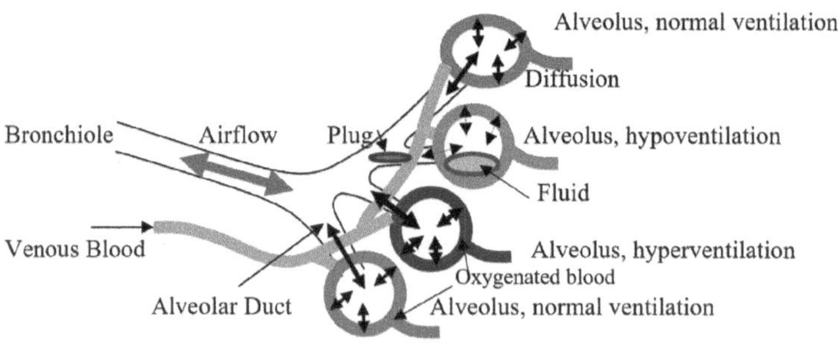

Fig. 9.3 The air moves in the bronchioles and alveoli. When the airflow stops in the alveolus, then diffusion through the respiratory membrane between the alveolus and the capillary takes over the gas exchange. If there is a plug in the airway passage to the alveolus and inflammatory exudate in the alveolus, the hypoventilation of the alveolus occurs. In pneumonia, the partial lung collapse occurs due to reduced distensibility of the lung as some alveoli fill with the exudate or pus, and the ventilation is impaired as oxygen is hard to reach the bloodstream due to a decrease in the diffusion capability in some alveoli.

(salbutamol) are used to increase airflow in conditions such as asthma and chronic obstructive pulmonary disease (COPD). Muscarinic antagonist medications such as ipratropium provide additional benefits when used in combination with albuterol in those with moderate or severe symptoms. In asthma, airflow limitation is generally fully reversible and caused by bronchoconstriction and secretion of mucus. Muscarinic antagonist, ipratropium bromide is effective for the relief of severe asthma attacks, when used in combination with albuterol, by blocking cholinergic tone and also by blocking reflex bronchoconstriction mediated by the vagus nerves. They may also inhibit secretion and enhance the clearance of mucus. The most common cause of the chronic obstructive pulmonary disease (COPD) is long-term tobacco smoking that causes an inflammatory response in the lungs, resulting in narrowing of the small airways, breakdown of the lung tissue, and finally emphysema. The diagnosis is based on the increased airway resistance with reduced airflow. In contrast to asthma, in COPD airflow limitation is not fully reversible, and is caused by structural changes and narrowing of peripheral airways along with parenchymal destruction. Muscarinic antagonists may increase airflow in COPD by blocking the cholinergic tone of the airway smooth muscle, but effects are limited. Stopping of smoking is required to improve the airflow. To overcome the airflow reduction, the extra pressure change between the external atmosphere and the alveoli should be created by deep forced breaths or mechanical inhalation.

Question 9.1 Why does the pressure in the room increases as the patient in a body plethysmograph inhales and the pressure decreases as the patient exhales?

Answer During inhaling, about 500 mL of the atmospheric air enters the patient's air passageways and the alveoli, but the thoracic wall expands more than 500 mL because the pressure in the lungs is lower than the pressure of the air in the body plethysmograph. The increased volume of the inhaled air pressures the air in the body plethysmograph.

9.6 Functional residual capacity and pulmonary compliance

Functional Residual Capacity (FRC) is not the residual volume under a forced breath. FRC is the volume of air present in the lungs at the end of normal passive expiration where the inward elastic recoil force of the lungs and outward expansion of the thoracic (chest) wall are in equilibrium and there is no exertion by the diaphragm or other respiratory muscles.

An elevated FRC may be an indication of emphysema since the reduced pulmonary compliance disturbs the equilibrium between the inward recoil of the lungs and the outward expansion of the chest wall. Patients with emphysema often have noticeably broader chests due to the reduced inward recoil force of the lungs. Obese patients will have a lower FRC in the supine position due to the increased inward recoil of the chest wall.

Pulmonary compliance is large if the volume of lungs expanding more against a given difference in transpulmonary pressure. Pulmonary compliance is an index of distensibility of the lungs. The compliance decreases in patients with chronic obstructive pulmonary disease such as chronic inflammation and emphysema due to tobacco smoking, or infections that cause fibrosis. Disturbance of surfactant (the alveolar phospholipid film) by the exudate or pus also decreases pulmonary compliance as seen in pulmonary edema and pneumonia.

Question 9.2 (1) Calculate transpulmonary pressure using the values of intrapleural pressure (p_{ip}) in Fig. 9.2 that we can measure experimentally. Transpulmonary pressure is the difference between intrapulmonary pressure (p_{pul}) and intrapleural pressure. (2) Tidal volume is the amount of air that moves in or out of the lungs during normal quiet breathing. Calculate pulmonary compliance (C_L) referring to Fig. 9.2 when tidal volume is 500 mL. $C_L = \frac{\Delta V_L}{\Delta (p_{pul} - p_{ip})}$ where $\Delta (p_{pul} - p_{ip})$ is the transpulmonary pressure change and ΔV_L is the volume change of the lungs at two points that are end-expiratory position and end-inspiratory position.

Answer (1) When the gas flow is zero, intrapulmonary pressure is supposed to be equal to atmospheric pressure. The transpulmonary pressure is (760–756) = 4 mmHg at the end-expiratory position and (760–754) = 6 mmHg at the end-inspiratory position. (2) $C_L = \frac{\Delta V_L}{\Delta (p_{pul} - p_{ip})} = \frac{500\ \text{mL}}{(6-4)\ \text{mmHg}} = 250\ \text{mL/mmHg}$.

9.7 Pulmonary resistance and compliance during inhalation of 5% oxygen

Fluorocarbon is a compound containing carbon and fluorine used chiefly as refrigerants, and formerly aerosol propellants. Trichlorofluoromethane (FC11) causes slight bronchoconstriction probably stimulating muscarinic receptors. Halothane, a fluorinated anesthetic administered in 4% concentration, also provokes an increase in pulmonary resistance. The prior administration of atropine prevents the onset of bronchoconstriction elicited from both gas exposures, indicating the vagus nerve mediates the airway response of these fluorinated gases [6].

In all, 5% oxygen is 25% of the normal oxygen supply. The percent O_2 saturation of Hb at this level of oxygen supply is estimated to be about 50% according to Fig. 9.1, which indicates exposure to 5% oxygen that will cause hypoxia. Lack of oxygen, increased carbon dioxide, and H^+ in arterial blood stimulate peripheral chemoreceptors at carotid and aortic bodies that send afferent impulses to the medullary respiratory centers to evoke activation of the respiratory muscles. The ventilation increases due to increased respiratory rate and increased absolute value of intrapleural pressure via reflexes initiated by the peripheral chemoreceptors. The additional contributors such as a decrease in pulmonary resistance associated with an increase in pulmonary compliance as shown in Fig. 9.4 are probably due to stimulation of β_2 receptors by circulating epinephrine (EPI) on the bronchial and bronchiolar smooth muscles and an increase in surfactant of the alveoli, indicating stimulation of the adrenal medulla mediated by the sympathetic nerve.

Question 9.3 Please list possible causes of hypoxia.

Answer (1) Fewer red blood cells, or too little hemoglobin contained in red blood cells. (2) Impaired blood circulation. (3) Histotoxic hypoxia when body cells cannot use oxygen as seen in cyanide poisoning. (4) Reduced arterial p_{O_2} as seen in pulmonary diseases such as pneumonia that impair ventilation. (5) Reduced arterial p_{O_2} as seen in exposure to the oxygen below 8% concentration.

9.8 Pneumonia

Pneumonia is an inflammation of the lung affecting primarily the small air sacs known as alveoli. In pneumonia, the partial lung collapse occurs due to reduced distensibility of the lung as some air sacs fill with inflammatory exudate or pus, and the fused basement membranes are separated by the interstitial exudate. The ventilation is impaired in the pathological states as oxygen is hard to reach the bloodstream due to a decrease in the diffusion capability in some alveoli. Typically, symptoms include some combination of sputum, cough, chest pain, fever, the difficulty of breathing, and high respiratory rate. Pneumonia is usually caused by infection with viruses or bacteria. Factors predisposed to pneumonia include chronic obstructive pulmonary disease, asthma, diabetes, heart failure, a history of smoking, and cystic fibrosis. Diagnosis is often based on the symptoms and physical

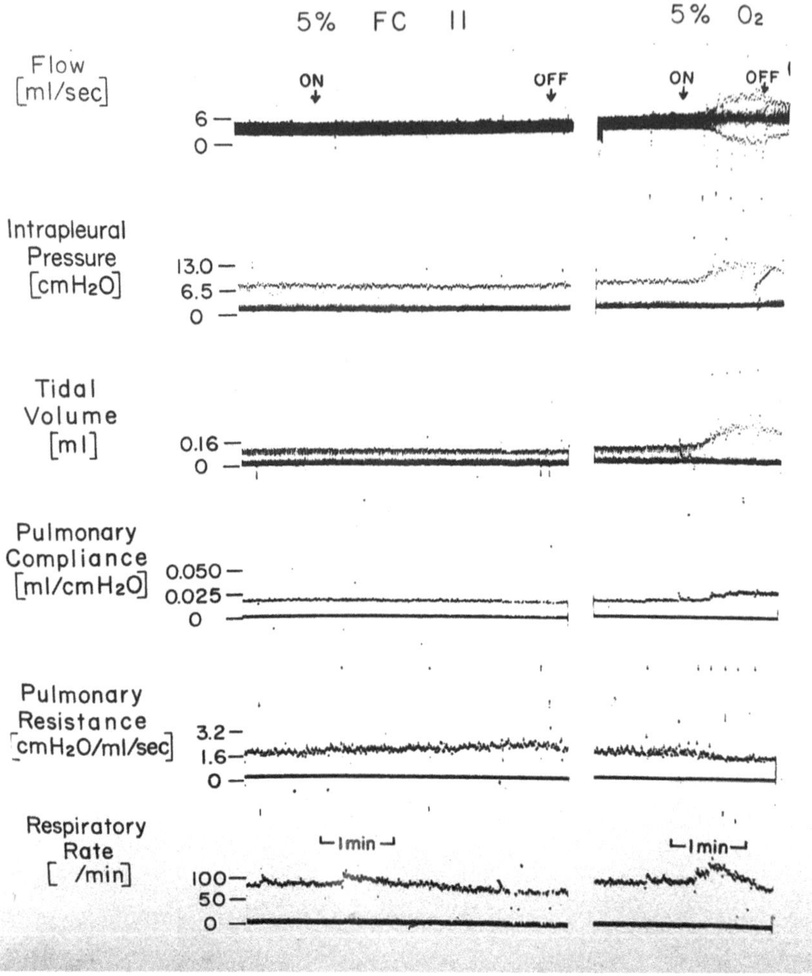

Fig. 9.4 Male Swiss mice were anesthetized by an intraperitoneal injection of sodium pentobarbital (65 mg/Kg). A plastic catheter was inserted into the trachea via a tracheostomy for exposure to 5% trichlorofluoromethane (FC11) and 5% oxygen. The mice were placed in the body plethysmograph to measure the tracheal airflow by a pneumotachograph inserted into the end of the plethysmograph instead of the tracheal cannula. Intrapleural pressure was measured by a differential manometer with one side attached to the atmosphere and the other to a catheter that had been inserted into the pleural cavity via a surgical incision and a puncture. Tidal volume was determined by the integration of the flow pattern [6,7].

examination. The patchy consolidations of the part of the lobe in chest X-ray would confirm the diagnosis with the help of blood tests and the examination of the sputum.

Bacteria are the most common cause of community-acquired pneumonia with streptococcus pneumonia isolated in nearly 50% of cases [8]. A number of drug-resistant versions of the above infections are becoming more common, which includes drug-resistant streptococcus pneumonia and methicillin-resistant *Staphylococcus aureus* (MRSA). In adults, *viruses* account for approximately a third and in children for about 15% of pneumonia cases. Commonly implicated agents include rhinoviruses, coronaviruses, influenza viruses, and so on [9]. Different viruses predominate at different periods of the year, for example, during flu season influenza viruses may account for over half of all viral cases. Outbreaks of other viruses also occasionally occur, including hantavirus and coronavirus. The COVID-19, SARS, and MERS are caused by coronaviruses which belong to a large family of single-stranded RNA viruses that can be isolated in different animal species [10]. Coronaviruses can cause multiple system infections in various animals and mainly respiratory tract infections in humans, as seen in severe acute respiratory syndrome (SARS) and the Middle East respiratory syndrome (MERS) [11]. Most patients with COVID-19 have mild symptoms and a good prognosis so far, but a few patients have developed severe pneumonia, pulmonary edema or multiple organ failure, and have died. Mechanical ventilation is required for the patients with severe respiratory insufficiency due to pneumonia with edema. The coronavirus does not seem to predominate at different seasons of the year so far. The blood plasma containing a large number of antibodies taken from the cured patient and antiviral agents such as remdesivir, molnupiravil and paxlovid, when administered in the early phase of the infection, were reportedly effective on pneumonia in some patients with COVID-19. The antibody test for the coronavirus could identify patients who have been infected with COVID-19, a key for knowing who's immune and for developing a vaccine. The antibody test using blood drawn from a vein takes 15 min and may give information of the prior experience of infection with the SARS-CoV-2 virus. The reverse transcription polymerase chain reaction (RT-PCR) test (see Section 3.11) reveals the presence of the virus in the body. Using saliva as a sample is now standard since it is easier than taking a mucus sample from the upper pharynx. *The salivary viral load was highest during the first week after symptom onset and subsequently declined with time.* The ratio of the mean virus amount ($\log_{10}$ copy/mL) in the posterior oropharyngeal saliva sample to endotracheal aspirate was found to be about

80% [12]. From the 74 COVID-19 patients, 323 serum samples, 247 urine samples, and 129 stool samples were collected and tested. The SARS-CoV-2 RNA detection rates were 2.8% (9/323), 0.8% (2/247), and 10.1% (13/129), and the mean viral load was 1210 ± 1861, 79 ± 30, and 3176 ± 7208 copy/μL, respectively [13]. The results indicate 10% of the patients have intestine infections besides the pharynx and the lung with the virus, and 2.8% of the patients were suspected to have platelet deficiency probably due to immune system malfunction since oral administration of dexamethasone is effective for releasing the symptoms. The risk of viral transmission via urine is expected to be low since the detection rate and amount is low. The antigen test using saliva as a sample is now available to see the presence of the virus in the body in hours though it is less sensitive to detect.

9.9 Conclusions

Pulmonary compliance is an index of distensibility of the lungs. The compliance decreases in patients with chronic obstructive pulmonary disease (COPD) such as chronic inflammation and emphysema due to tobacco smoking, or infections that cause fibrosis. Disturbance of alveolar phospholipid film by the exudate or pus also decreases pulmonary compliance as seen in pulmonary edema and pneumonia.

In COPD, airflow limitation is not fully reversible and is caused by structural changes and narrowing of peripheral airways along with parenchymal destruction. Muscarinic antagonist medications such as ipratropium provide additional benefits when used in combination with short-acting β_2 adrenergic agonists, albuterol, but the effects are limited.

During hypoxia, the ventilation increases due to increased respiratory rate and intrapleural pressure resulting from stimulation of the respiratory muscles. Additionally, a decrease in pulmonary resistance and increase in pulmonary compliance also contribute to an increase in ventilation, probably due to stimulation of β_2 receptors by circulating epinephrine (EPI) on the bronchial and bronchiolar smooth muscles and an increase in surfactant of the alveoli, indicating stimulation of the adrenal medulla mediated by the sympathetic nerve.

In pneumonia, the partial lung collapse occurs due to the reduced distensibility of the lung when some air sacs fill with inflammatory exudate or pus, and the fused basement membranes are separated by interstitial exudate. The ventilation is impaired in the pathological states as oxygen is hard to reach the bloodstream due to a decrease in the diffusion capability in some alveoli.

References

[1] G.W. Castellan, Physical Chemistry, third ed., Addison-Wesley, Reading, MA, 1983 (Chapter 14).
[2] E.N. Marieb, K. Hoehn, The respiratory system, in: Human Anatomy & Physiology, eighth ed., Benjamin Cumings Pearson, San Francisco, CA, 2010 (Chapter 22).
[3] S.M. Johnson, J.C. Smith, G.D. Funk, J.L. Feldman, Pacemaker behavior of respiratory neurons in medullary slices from neonatal rat, J. Neurophysiol. 72 (6) (1994) 2598–2608.
[4] J.H. Comroe Jr., R.E. Foster, A.B. Dubois, W.A. Briscoe, E. Carlsen, Diffusion, in: The Lung: Clinical Physiology and Pulmonary Function Tests, second ed., Year Book Medical Publishers, Chicago, IL, 1963 (Chapter 5).
[5] J.H. Comroe Jr., R.E. Foster, A.B. Dubois, W.A. Briscoe, E. Carlsen, Mechanics of breathing, in: The Lung: Clinical Physiology and Pulmonary Function Tests, second ed., Year Book Medical Publishers, Chicago, IL, 1963 (Chapter 7).
[6] R.S. Brody, T. Watanabe, D.M. Aviado, Toxicity of aerosol propellants on the respiratory and circulatory systems, Toxicology 2 (1974) 173–184.
[7] D.M. Aviado, T. Watanabe, Functional and biochemical effects on the lung following inhalation of cigarette smoke and constituents. 1 High- and low-nicotine cigarettes in mice, Toxicol. Appl. Pharmacol. 30 (1974) 185–200.
[8] S. Anevlavis, D. Bouros, Community acquired bacterial pneumonia, Expert Opin. Pharmacother. 11 (3) (2010) 361–374.
[9] O. Ruuskanen, E. Lahti, L.C. Jennings, D.R. Murdoch, Viral pneumonia, Lancet 377 (9773) (2011) 1264–1275.
[10] M. Cascella, M. Rajnik, A. Cuomo, S.C. Dulebohn, D.N. Raffaela, Features, Evaluation and Treatment Coronavirus (COVID-19), StatPearls Publishing, 2020.
[11] N. Chen, et al., Epidemiological and clinical characteristics of 99 cases of 2019 novel coronavirus pneumonia in Wuhan, China: a descriptive study, Lancet 395 (2020) 507–513.
[12] K.K. To, O.T. Tsang, W. Leung, A.R. Tam, T. Wu, D.C. Lung, et al., Temporal profiles of viral load in posterior oropharyngeal saliva samples and serum antibody responses during infection by SARS-CoV-2: an observational cohort study, Lancet Infect. Dis. (2020), https://doi.org/10.1016/S1473-3099(20)30196-1.
[13] J.M. Kim, H.M. Kim, E.J. Lee, H.J. Jo, Y. Yoon, N.J. Lee, J. Son, Y.J. Lee, M.S. Kim, Y.P. Lee, et al., Detection and isolation of SARS-CoV-2 in serum, urine, and stool specimens of COVID-19 patients from the Republic of Korea, Osong Public Health Res. Perspect. 11 (3) (2020) 112–117.

CHAPTER 10

Electrochemical mechanism of arrhythmias and antiarrhythmic drugs

10.1 Impulse conduction in the heart

In normal sequence, cyclic cardiac impulses originate in the SA node which has pale oval primitive pacemaker cells in the deeper portion. Once impulse leaves the sinoatrial (SA) node through the gap junction, it spreads rapidly in the atria and transmits to the atrioventricular (AV) node through the internodal and interatrial pathways. From the AV node, a bundle of His takes over the impulse transmission, and then through Purkinje fibers, it spreads in the ventricular muscles. The thin filaments called Purknje fibers distributing to ventricular contractile cells conduct impulses well, and the speed is the fastest among the cardiac cells since they are thin filaments having many voltage-gated Na^+ channels in the membrane-like neurons. Its velocity of conduction is 4 m/s. Conduction in the bundle of His and contractile myocardiac cells is not as fast as the Purkinje fibers, and their velocities are about 1 m/s. However, the conduction rate of the AV node is the slowest (0.05 m/s) since the AV node is made up of a mixture of the most frequent slender transitional cells and less frequent primitive pacemaker cells in its deeper portions which lack voltage-gated Na^+ channels, where diffusion of Ca^{2+} through gap junctional channels is thought to be the main contribution to transmission of the impulse. The slow conduction in the AV node is a prerequisite to give the atrial contractile muscle fibers a time to propel all the blood into the ventricle before the ventricles contract. The AV node, the bundle of His, and Purkinje fibers are called the latent pacemaker. Each time the SA node fires, it resets the slower latent pacemakers. If the SA node stopped firing, then the AV node would take over at a slightly slower pace. Most of the ectopic small depolarizations in a resting state between −70 and −55 mV are thought to be removed by an outward current of IRK channel by means of removal of the plug from the channel pore and wither without generating an action potential (see Section 7.2). In the heart, sympathetic

Modeling Electrochemical Dynamics and Signaling Mechanisms in Excitable Cells with Pathological Case Studies
https://doi.org/10.1016/B978-0-323-98803-2.00013-4

stimulation can raise cAMP and Ca^{2+} levels in myocardial cells and increase the excitability of the cells by activating the actin-activated ATPase on the myosin head. If ATPase in the myosin head becomes more active at a lower concentration of Ca^{2+}, the muscle fibers would get more excitable and induce arrhythmias (see Section 8.4). Disfunction of IRK and high cytosolic Ca^{2+} supposedly predispose arrhythmias.

The P wave of the electrocardiogram (ECG) corresponds to the depolarization of the myocardium in the atria that elicits the atrial systole. No P wave may be observed in the atrial fibrillation. The QRS complex on the ECG relates to the depolarization of the myocardium throughout the ventricles, causing ventricular contraction. The T wave represents ventricular repolarization. The empty space called the PQ segment corresponds to the pause at the AV node. The ST segment corresponds to the interval between ventricular depolarization and repolarization. The TP segment represents the heart's quiet time when the cells finish repolarization and are ready for another signal (Fig. 10.1). Decreased AV conduction may increase PQ segment, and increased AV conduction may shorten the PQ segment. The rate of action potential in the SA node which normally determines the heart rate is controlled by the postganglionic autonomic nerves. The sympathetic stimulation accelerates the heart rate and parasympathetic stimulation slows the heart rate.

Transmission may be disturbed if an ectopic focus in the heart fires a depolarization wave during or right after repolarization of the preceding impulse from the SA node. If an ectopic site is in the ventricles, the SA node that is out of refractory can fire off a pulse without any interference, and the atria can contract right on schedule, but the pulse cannot travel the ventricles in the refractory period, and the ventricles cannot contract. There arises a compensatory pause between ventricular contractions, bringing more ventricular filling for a strong contraction.

10.2 Disorder of impulse generation and ECG

Arrhythmias occur if the disorganized impulse originating in ectopic focuses somewhere in the heart interferes with normal transmission of an impulse from the SA node or causes abnormal extra loop circuits. If there is an extra anatomical or pathological circuit inside the heart, it may act as a short circuit and make the heartbeat abnormally fast and irregular.

Premature atrium contraction (PAC) are premature heartbeats originating in the atria. In most cases, conduction through the AV node and the ventricles

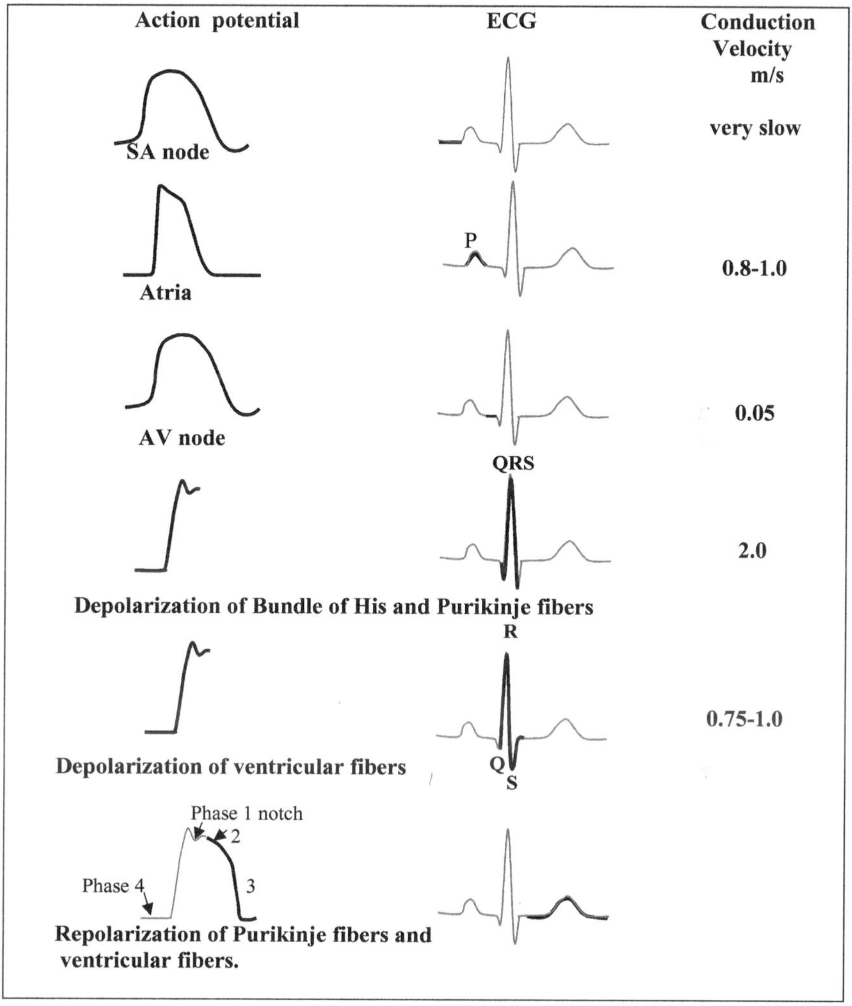

Fig. 10.1 Action potential, electrocardiogram, and conduction velocity of cardiac cells from different regions of the heart. The P wave of the electrocardiogram corresponds to the depolarization of the myocardium in the atria that elicits the atrial systole. Repolarization in the atrial myocardium cannot be seen due to the large depolarization of the ventricular fibers. The empty space called the PQ segment corresponds to the pause at the AV node. Depolarization of Purkinje fibers and ventricular fibers occur about the same time corresponding to the QRS complex. The T wave represents ventricular repolarization. Purkinje cells specialized for the rapid conduction are distributed throughout the inner ventricular walls of the heart and propagate the impulse 2 times much faster than any other contractile myocardial cells since they have denser Na^+ channels than any other tissues.

is not affected. Most PACs elicit irregular heartbeats with the normal shape of the QRST complex. An ectopic atrial activation can usually enter the SA node, depolarize it, and reset its timing, causing a so-called noncompensatory pause, that is, the PP intervals that contain the ectopic beats is less than two times of normal PP interval. The exact cause of PAC is unclear. PACs may be worsened by caffeine and alcohol. PACs are common among patients with lung disorders such as chronic obstructive pulmonary disease, but they also occur in healthy young and elderly people. PACs are often completely asymptomatic. In most cases, no treatment is needed for PACs, although medications such as beta blockers can reduce the frequency of symptomatic PACs (Fig. 10.2).

Premature ventricular contraction (PVC) is a relatively common event. A ventricular ectopic focus somewhere in the ventricles sends an early depolarization wave before the SA node gets fire. Depolarization originates in the ventricles rather than coming from the pacemaker cells in the SA nodes. While PACs elicit the normal shape of the QRST complex, PVCs show the abnormal shape of the QRST complex. When an ectopic focus originates in the left ventricle and moves toward the right ventricle, the V1 lead shows a large positive complex with a dominating R wave. When an ectopic focus originates in the right ventricle and then moves toward the left ventricle, the V1 lead shows a large negative complex with a dominating S wave. Single PVC may cause no symptoms. However, very frequent (over 20% of all heartbeats) PVCs are considered as a risk factor for arrhythmia-induced cardiomyopathy [1] (Fig. 10.2).

Atrial fibrillation is characterized by the absence of P waves and irregular intervals of beats. The disorganized impulses originate in ectopic sites anywhere in and around the atria, commonly near the roots of pulmonary veins. The pulmonary veins are reported to have unique electrophysiological properties triggering and maintaining atrial fibrillation. The unsynchronized chaotic impulses cause the atria to quiver or fibrillate rather than contract. Most of the impulses do not pass through the AV node to the ventricles due to the refractory properties of the cells in the AV node. Those which come through are irregular. As the atria do not function properly, the atrium does not empty blood completely into the right ventricle. The blood may stagnate inside the atria, and blood clots may form. No specific treatment is typically required, though aspirin or anticlotting medication may be considered for prevention of the clotting of blood in the fibrillating atria (Fig. 10.2).

Ventricular fibrillation can occur due to coronary heart disease, valvular heart disease, or cardiomyopathy. Quivering waves of ECG are characteristic of

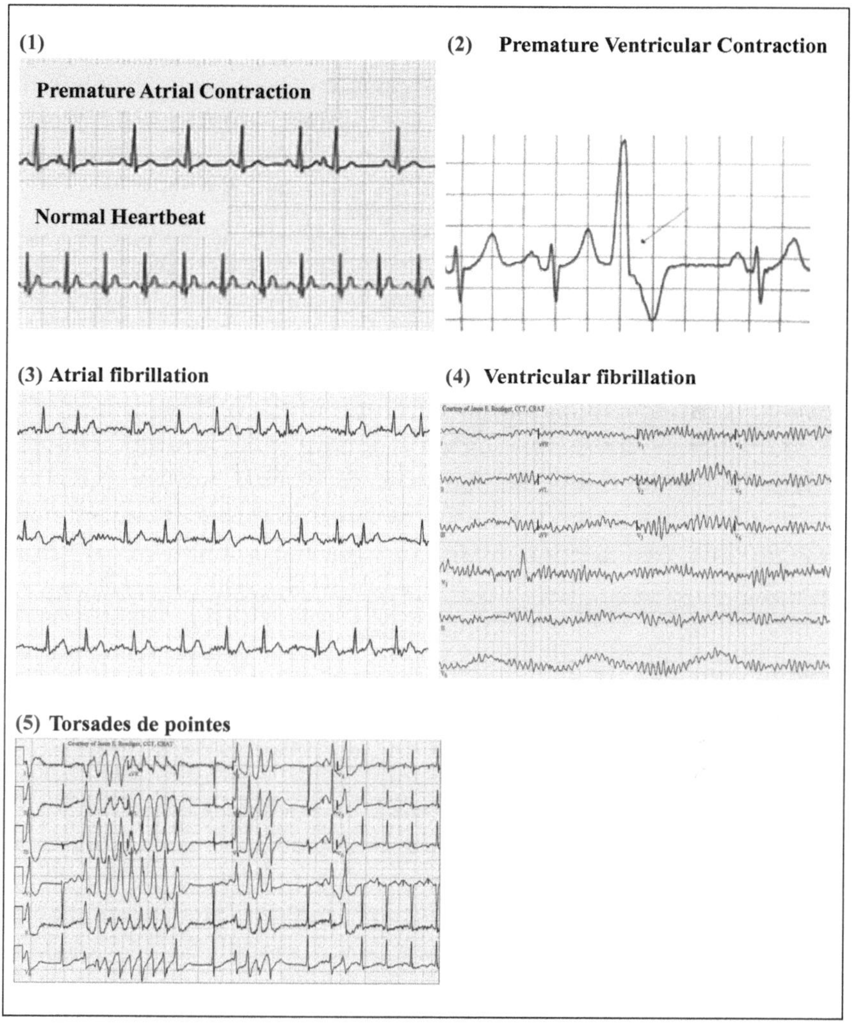

Fig. 10.2 (1) Premature atrium contraction elicits irregular heartbeats with the normal shape of the QRST complex. An early atrial activation causes the P wave to merge with the preceding T wave producing a peaked fusion wave. (2) Premature ventricular contraction (PVC) shows the abnormal shape of the QRST complex. The V1 lead shows a large positive complex with a dominating R wave, indicating the ectopic focus originates in the left ventricle and moves toward the right ventricle. (3) Atrial fibrillation is characterized by the absence of P waves and irregular intervals between heartbeats. (4) Ventricular fibrillation is disorganized contractions of various parts of the ventricle, and the heart quivers instead of pumping. (5) Torsades de pointes (twisting of peaks) is a specific type of polymorphic ventricular tachycardia that can lead to sudden cardiac death.

ventricular fibrillation caused by disorganized depolarizations in the various parts of the ventricles. Ventricles of the heart quiver instead of pumping normally. This is followed by cardiac arrest in the absence of treatment. The fibrillation is treated not only with voltage-gated Ca^{2+} channel blockers and β blockers but also with the application of a large electric current across the chest to reset by means of depolarizing all the muscle fibers at once (Fig. 10.2).

Torsades de pointes, "twisting of peaks" is a specific type of polymorphic ventricular tachycardia that can lead to sudden cardiac death. In case of marked prolongation of the action potential seen in congenital long QT syndrome, quinidine-induced QT prolongation, or after block of K^+ channels, an abnormally prolonged action potential may occur due to superimposition of two waves. It exhibits distinct characteristics on ECG. Prolongation of the QT interval can increase a person's risk of developing this abnormal heart rhythm. Common causes for Torsades de pointes include quinidine-induced QT prolongation particularly when serum magnesium and potassium are low (see Section 10.6), or congenital long QT syndrome (Fig. 10.2).

10.3 Incidence of premature contractions

Let us discuss when and how premature contraction occurs in the heart. A high level of cytosolic calcium ion is known to make the cardiac muscle excitable by means of activating the actin-activated ATPase on the myosin head, which can be a cause of premature contractions along with disfunction of inwardly rectifying (IRK) currents.

Sympathomimetic activation, stress and cardiac ischemia were reported to induce premature myocardial contractions. Cytosolic Ca^{2+} overload has been reported [2] after sympathomimetic stimulation and during cardiac ischemia. The mechanism of the incidence of Ca^{2+} overload can be explained as follows.

First, phosphorylation of RyR2 by protein kinase A activated by sympathomimetic stimulation may make the RyR2 Ca^{2+} channel leakier [3], which aggravates Ca^{2+} leak from the SR and results in the Ca^{2+} overload. In vitro, PKA can phosphorylate RyR2. However, in intact myocytes, a functional role for PKA regulation of RyR2 remains controversial [4]. Some evidence is in favor of Ca/calmodulin-dependent protein kinase II (CaMKII) as the prime modulator of RyR2 during β-adrenergic stimulation. β-Adrenergic receptor agonist, e.g., isoproterenol enhanced Ca^{2+} leak from the SR *in intact* rabbit ventricular myocytes, and the Ca^{2+} leak seemed not to be related to protein kinase A activity but was thought to be mediated by CaMKII activity [5,6]. It is considered that an increase in endogenous NO production in cardiomyocytes by β-adrenergic stimulation results in the NO-dependent activation of CaMKII and a subsequent increase in diastolic Ca^{2+} discharge [7].

Secondly, an increase in the cytosolic cyclic AMP level by sympathomimetic stimulation may cause an increase in the cytosolic Ca^{2+}.

Thirdly, ATP-deficiency during ischemia may disturb the pumping of Ca^{2+} out of the cytosol by active transporters such as sarcolemma Ca^{2+} ATPase (PMCA) and sarcoplasmic reticulum Ca^{2+} ATPase (SERCA), which will contribute the further increase in the cytosolic Ca^{2+}.

Fourthly, the intracellular increase in Ca^{2+} by sympathomimetic stimulation causes the activation of Na^+-Ca^{2+} exchanger which removes Ca^{2+} if the intracellular Na^+ is not high and ATP is sufficient. But in the case of ischemia, the Ca^{2+} load of the cell becomes excessive due to the reverse action of Na^+-Ca^{2+} because of the high level of cytosolic Na^+ raised by dysfunction of the Na^+-K^+ pump under the ATP deficiency.

An increase in the intracellular Ca^{2+} would excite the cardiac cell to contract by the following reasons. First, the cell will be more easily to be depolarized if a level of Ca^{2+} increases in a resting state. Second, higher Ca^{2+} activates myosin ATPase, and the hydrolysis of ATP by the activated myosin ATPase causes myosin head to bind to a high energy phosphate and a re-cocking of the myosin head to ready for rebinding to actin, which enables cardiac muscle fibers to contract prematurely (see Section 8.4). Contrary to that IRK channels play an important role in setting and stabilizing the resting potential and thus preventing the heart from arrhythmias. Sympathomimetic activation and stress can deteriorate the function of IRK currents. An increase in cyclic AMP level as seen in sympathomimetic stimulation causes inhibition of inwardly rectifying K^+ currents along with an increase in Ca^{2+} in the myocardium, which probably increases an incident of premature contractions (see Section 7.2).

The decrease in contractile force under ATP deficiency and uncoordinated contractions under Ca^{2+} overload may be a sign of sudden cardiac arrest and appears to be associated with a decrease in high-energy phosphates. To recover, the premature depolarizations should be treated with administrations of an adequate amount of Ca^{2+} channel blockers, and or β blockers as well as oxygen supply.

10.4 Cardiac arrhythmias

The normal impulse transmission will be disturbed if ectopic focuses somewhere in and around the atria fire disorganized impulses as seen in atrial fibrillation, or if an ectopic focus in the heart fires an action potential during the repolarization phase as seen in Torsades de pointes or just after the preceding T wave of the pulse from the SA node as seen in premature contractions, or if there is an extra electric conduction route where there are cases that a loop pathway sometimes happens to be formed by a premature action potential originating in an ectopic focus as seen in supraventricular tachycardias. Supraventricular tachycardia is defined as an abnormally fast heartbeat above the ventricles.

Ventricular tachyarrhythmia occurs when a transient loop circuit around something like scar tissue in the ventricles is formed under an acute pathological condition (see Section 10.4.3). The Ca^{2+} overload, valvular heart disease, or cardiomyopathy would predispose the ventricular tachyarrhythmia to ventricular fibrillation.

In addition to sympathetic stimulation, high intracellular Na^{+} and ATP-deficiency will precede Ca^{2+} overload as seen in coronary heart disease. Low conductivity of the cells and high Ca^{2+} as seen in ischemic damage or under sympathetic stimulation or low K^{+} seen in hypokalemia may predispose latent pacemakers such as the AV node, the bundle of His, and Purkinje fibers, or cardiac muscle cells to get enhanced automaticity and fire a premature action potential before the impulse coming from the SA node. The cell with high Ca^{2+} under sympathetic stimulation or ischemic condition is considered as a candidate for an ectopic focus of the premature contractions since it has not only low ion conductivity but also is excitable.

The premature contractions induced by sympathetic stimulation may be reduced by the administration of an adequate amount of β blockers, and supraventricular tachycardia can be controlled by Ca^{2+} channel blockers and β blockers (see Section 10.6).

10.4.1 Bradycardia and tachycardia

When the membrane potential of the pacemaker cells reaches the threshold (−50 to −60 mV), plasma membrane L-type voltage-gated Ca^{2+} channels open and generate an action potential. The pace of the pulse gets faster when the second messenger, i.e., cAMP production is activated by the sympathetic nerve while it gets slower when the G protein-gated inwardly rectifying potassium channels (IKACh) is activated by the vagus nerve. The cytosolic cAMP raised by the sympathetic stimulation causes tachycardia and is considered arrhythmic (see Section 7.9). However, parasympathetic activation causes bradycardia and is considered antiarrhythmic [8].

High calcium in the cell is very important for the cardiac muscle to contract since activated myosin ATPase in the presence of Ca^{2+} causes the myosin head to bind to a high-energy phosphate, which enables the head to recock. β-Adrenergic receptor agonist, i.e., isoproterenol enhanced Ca^{2+} leak from the SR in intact rabbit ventricular myocytes. The Ca^{2+} leak seemed not to be related to protein kinase A activity but was thought to be mediated by CaMKII activity [5] (see Section 7.4). Through inactivation of inwardly rectifying K^+ currents and an increase in Ca^{2+}, the sympathetic stimulation is implicated in making cardiac muscles to be excitable.

Since there is an influx of Ca^{2+} through voltage-gated L-type Ca^{2+} channel in the sarcolemma and a massive discharge from the SR during systole, the plasma membrane Ca^{2+} ATPase (PMCA), sarcoendoplasmic reticulum Ca^{2+} ATPase (SERCA), and Na^+-Ca^{2+} exchanger are considered to be a critical player in removing Ca^{2+} during diastole in order to make cardiac contraction cycle (see Section 7.5). The failure to recover from an increase in cytosolic Ca^{2+} concentration may result in increased action potential duration and slow pulse rate as seen in digoxin side effects. ATP-deficiency in a poor oxygen supply preceding the cytosolic Ca^{2+} overload will make cardiac muscles to be more excitable and predisposed to tachyarrhythmias. Tachyarrhythmias can be also induced by the formation of anatomical or pathological loop circuits (see Section 10.4.3).

Some drugs have been known to induce bradycardia or tachycardia. Bradycardia may occur when vagal tone is increased by ACh or digoxin side effect, or when the sympathetic tone is reduced after administration of β blockers such as propranolol, sotalol, and celiprolol in chronic therapy of hypertension, or when the Ca^{2+} channels are blocked by verapamil, or when the propagation of impulse is suppressed by local anesthetics. Tachycardia may occur when sympathomimetic drugs are administered. Withdrawal of β blockers used in chronic therapy of hypertension may cause tachycardia since more β adrenergic receptors are available for β receptor agonists after the withdrawal.

Example 10.1 Describe the action mechanism of digoxin, and treatments of bradycardia induced by digoxin side effects in case of sufficient ATP and oxygen.

Binding of digoxin to Na^+-K^+ ATPase inhibits its activity mainly in the myocardium. This inhibition causes an increase in intracellular sodium ion level, resulting in a reversal action of Na^+-Ca^{2+} exchanger. The reversal exchange of the ions causes an increase in the intracellular calcium concentration. If ATP is sufficient, an increase in cytosolic Ca^{2+} leads to an increase in the storage of calcium in the SR, causing a corresponding increase in the release of calcium during systole. This leads to the increased inotropic action. The delayed recovery from an increase in cytosolic Ca^{2+} may result in increased duration of the repolarization phase of the action potential which gives the basis for bradycardia. Digoxin also causes prominent vagotonic actions particularly on the AV node, leading to bradycardia [9]. The bradycardia induced by digoxin will be treated by the administration of antidigoxin antibodies and temporary pacing.

10.4.2 Afterdepolarizations

A ventricular ectopic focus somewhere in the ventricles, that is a cell or area of tissue, sends an early depolarization wave before the SA node gets fire. Depolarization originates in the ventricles rather than coming from the pacemaker cells in the SA nodes. High Ca^{2+}, low PH, ATP-deficiency as seen in ischemic damage or under sympathetic stimulation may predispose latent pacemaker cells in the AV node, the bundle of His, and Purkinje fibers, or cardiac muscle cells to get enhanced automaticity and depolarize before the pacemaker cells in the SA node. Under sympathetic stimulation or ischemic condition, low conductivity of the cells, and high Ca^{2+} by activation of ryanodine receptor 2 channels, might predispose cardiac cells to get automaticity (see Sections 7.4 and 7.7).

Early after depolarization (EAD) is a premature ventricular depolarization that occurs during phase 3 corresponding to ventricular repolarization (T wave). The ectopic impulse is superimposed on the normal action potential from the pacemaker cells in the SA nodes. EADs occur in case of marked prolongation of the action potential seen in congenital long QT syndrome, quinidine-induced QT prolongation or after block of K^+ channels. EADs are consistently elicited during slow heart rate particularly when K^+ in the blood serum is low (hypokalemia). If EAD arises in the repolarizing phase of a normal action potential, an abnormally prolonged action potential called Torsades de pointes will occur due to the superimposition of two action potentials as shown in Fig. 10.2 (5), Fig. 10.3 (left), and Fig. 10.4 (right). EADs are commonly seen in Purkinje fibers and ventricular muscle fibers.

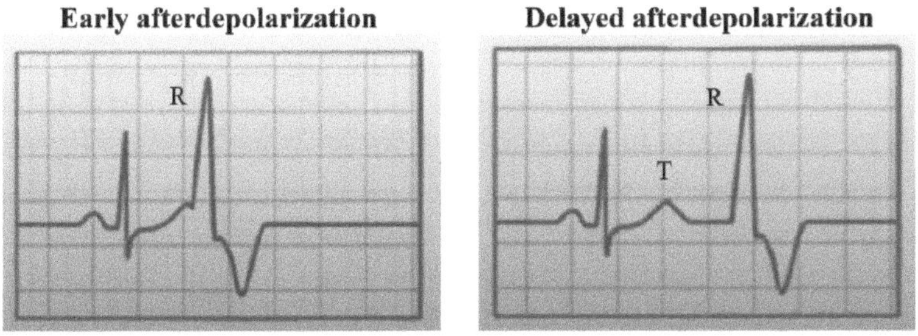

Fig. 10.3 Ventricular ectopic focus somewhere in the ventricles, that is a cell or area of tissue, sends a premature depolarization wave before the SA node gets fire. (1) When a premature depolarization in a cardiac cell happens during ventricular repolarization, it is called early afterdepolarization. A premature ventricular depolarization is superimposed on the preceding T wave of the beat initiated by the pacemaker cells in the SA nodes. (2) If it happens after repolarization is finished, it is called a delayed afterdepolarization. The V1 lead shows a large positive complex with a dominating R wave indicating that the ectopic focus originates in the left ventricle and moves toward the right ventricle.

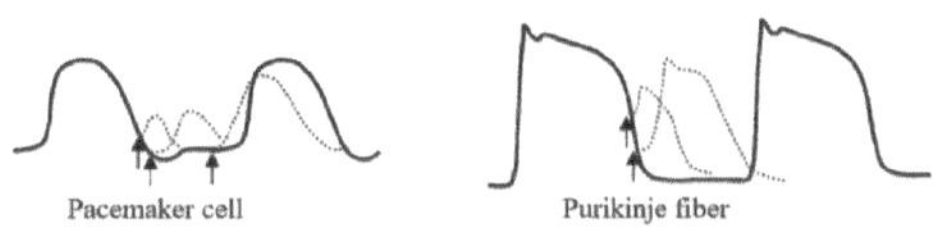

Fig. 10.4 Action potentials of the pacemaker cell and Purkinje fiber respond to premature stimuli *(arrow)* differently. Ion conductivity of pacemaker cells is small. Pacemaker cells conduct impulses in a slow manner. Refractory period of the pacemaker cells is shorter, but the magnitude of refractoriness depends on time. Ion conductivity of Purkinje cells is large. Purkinje cells conduct impulses in a fast manner. Refractory period of Purkinje fiber is longer, and the magnitude of its refractoriness depends on the number of Na^+ channels recovered from the inactive state. In a refractory period, impulse cannot travel in the cell. The degree of recovery of Na^+ channels from the inactive state can be altered by Na^+ channel blockers such as local anesthetics, which cause a longer refractory period. The premature stimuli during the repolarization phase of the preceding impulse from the SA node can be interfered with, depending on the degree of recovery of the channels from the inactive state. A premature action potential before completion of normal ventricular repolarization can be superimposed on the preceding pulse from the pacemaker cells in the SA node.

Delayed afterdepolarization (DAD) is a premature depolarization that occurs during phase 4 just after preceding repolarization is finished. So, the ectopic impulse is not superimposed on the preceding action potential originating in the SA node. Lead V1 on an ECG measures a depolarization wave that moves toward the right ventricle. When an ectopic focus originates in the left ventricle and moves toward the right ventricle, the V1 lead shows a large positive complex with a dominating R wave. When an ectopic focus originates in the right ventricle and then moves toward the left ventricle, the V1 lead shows a large negative complex with a dominating S wave. Regardless of the origin in the ventricle, a premature ventricular contraction often has an abnormal T wave since the timing and direction of repolarization will be abnormal as well.

10.4.3 Tachycardia induced by anatomical or pathological extra loop circuits

When an impulse can transmit through more than one pathway in the heart, a loop circuit may be formed by a pulse originating in an ectopic focus facilitating tachyarrhythmias. If there is an extra electric conduction route inside the heart, it sometimes acts as a short circuit, making the heartbeat abnormally fast. There are anatomical abnormal extra circuits or pathological extra circuits.

Anatomically abnormal extra pathway: Patients with Wolff-Parkinson-White syndrome have an anatomically abnormal accessory pathway called the bundle of Kent (fast-conduction tissue) between the atrium and ventricle. An action potential generated in the SA node can enter the ventricles through both the AV node and the bundle of Kent. An action potential that enters the left ventricle through the bundle of Kent commonly stimulates the left ventricle cells shortly after the action potential enters the AV node (slow-conduction tissue). The characteristic EKG shows a shorter PQ interval and the slow rise in the QRS complex is called a delta wave. This Wolff-Parkinson-White pattern seems to cause no symptoms, but sometimes facilitates tachyarrhythmias leading to cardiogenic shock under rapid atrial beats $(250-300/\mathrm{min})$ if a closed circuit between the atria and the ventricles happens to be formed. It happens when a premature impulse enters the AV node, and the impulse cannot travel through the accessory pathway which happens to be in a refractory period.

Anatomical extra loop circuits: (a) Atrioventricular reentrant tachycardia (AVRT) has a reentry circuit with the normal atrioventricular conduction system and the accessory pathway which happens to be in a refractory period. A premature action potential that enters via the AV node spread over the ventricles while the signal cannot enter the accessory pathway in a refractory period. By the time the signal reaches the accessory pathway, the accessory pathway will no longer be in a refractory period. The signal can exit through the accessory pathway and enter the atrium, and then go back to stimulate the AV node. The signal can go around the reentry circuit.

Supraventricular tachycardia arises briefly between a period of normal heart rates, which is called paroxysmal supraventricular tachycardia. Most people with supraventricular tachycardia live healthy lives without restrictions or interventions. For others, treatment and lifestyle changes can often control or eliminate rapid heartbeats. Voltage-gated Ca^{2+} channel blockers have been administered to treat this syndrome to slow conduction through the AV node (see Section 10.6). Recurrences may be prevented by ablation of the extra circuit for permanent cure.

(b) An atrioventricular nodal reentrant tachycardia (AVNRT) involves two circuits, that is, the slow response cells in the center and the fast-response tissue around it in the AV node. There is no accessory pathway like in AVNT but there are two pathways that run through the AV node. One is a slow pathway containing oval primitive pacemaker cells where an impulse travels more slowly down the path through the gap junctional channels, and the other one around the slow pacemaker cells is *an anatomically abnormal fast pathway* containing slender cells with some voltage-gated Na^{+} channels where an impulse can travel faster through if it is not in a refractory period. A pathway separates into two pathways in the AV node and joins again after leaving the AV node. The slow pathway has a shorter refractory period while the fast pathway has a longer refractory period. When an impulse from the SA node has transmitted through the fast pathway and leaves the AV node, the fast pathway goes into a refractory period. When the signal passing through the slow pathway reaches the final common pathway, it hits the refractory period of both the fast pathway and the bundle of His and must terminate. When another impulse originating in the SA node comes to the AV node, both pathways are ready to accept it since they are recovered from their refractory period, and the impulse can pass the AV node faster through the fast pathway. However, if a premature impulse comes in when the fast pathway is still in a recovering phase from a refractory period, the impulse should go down the slow track. By the time the impulse reaches the final common pathway, the fast track is recovered from the refractory period. The impulse can travel back up the fast track, and it can enter the slow pathway as the slow track has been through its refractory period and recovered over time. The impulse can circle around continuously in the closed circuit. This supraventricular tachycardia has a narrow TQ interval with 200–250 beats per min. Patients with rapid heartbeats have been treated by the Ca^{2+} channel blockers to reduce the AV conduction.

Pathological extra loop circuits: Functionally defined reentrant ventricular arrhythmia is a type of tachycardia that occurs due to the formation of transient loop circuits around something like scar tissue in the ventricle under an acute pathological condition. The functionally defined reentry circuit is thought to include a combination of high resistive tissues as blocks of ionic currents and an area of low conductive tissue as an ectopic focus. The cell with high Ca^{2+} under sympathetic stimulation or ischemic condition is considered as a candidate for an ectopic focus of the premature impulse since it has not only low ion conductivity but also it is excitable. A sufficiently low conductive cell or area of tissue with high Ca^{2+}, accompanied by ATP-deficiency and high cAMP, can be an ectopic focus which might get enhanced automaticity and send a premature impulse. The scar tissue formed after a heart attack does not conduct ionic currents at all, i.e., the ionic conductivity of the scar tissue is almost zero (see Section 5.2), and it does not depolarize. The impulse cannot enter the scar tissues, but it possibly changes direction. It would go around and create a functionally defined reentrant closed circuit. If high resistive areas block the impulse, the impulse starts to curve, circles around instead of radiating and ends up generating another impulse at the original low conductive tissue (Fig. 10.5). The spiral may recycle or spread and be broken down into many spirals if there are multiple units of high resistive areas in different parts of the ventricles. Ventricular fibrillation following ventricular tachycardia may be caused by the multiple mini reentrant circuits within the ventricles. Ventricular fibrillation is the outcome of disorganized contractions in various parts of the cardiac muscle particularly found in patients with a heart valve disease or history of a heart attack. The fibrillation is treated not only with voltage-gated Ca^{2+} channel blockers and or β blockers but also with the application of a large electric

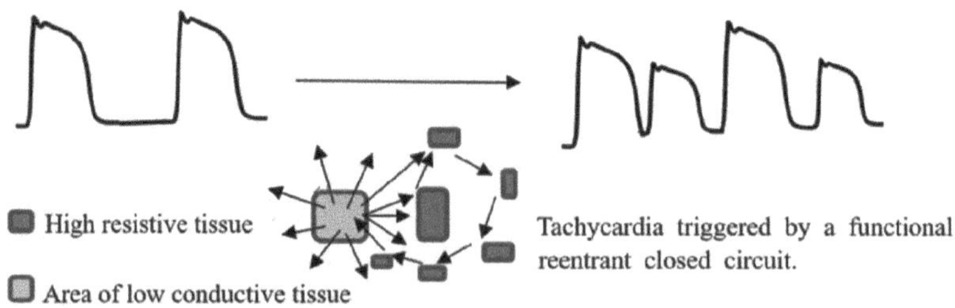

Fig. 10.5 The cell with high Ca^{2+} under sympathetic stimulation or ischemic condition is considered as a candidate for an ectopic focus of the premature impulse since it has not only low ion conductivity but also it is excitable. The scar tissues formed after a heart attack do not conduct ionic currents at all, block any impulses, and change the directions. The impulse starts to curve, circles around the scar tissue and ends up generating the next impulse in the original low conductive tissue, which makes a reentrant closed circuit.

current across the chest to reset by means of depolarizing all the muscle fibers at once. Implantable cardioverter-defibrillators, devices that can detect ventricular fibrillation and automatically delivering a defibrillating shock, have been used in patients with a high risk of ventricular fibrillation.

10.5 Experimental induction of ventricular tachycardia

Traumatic stress or tissue damage besides hypoxia can also change electrochemical properties of cardiac myocytes, which was examined by the inhalation of experimental emphysematous unanesthetized rat to trichloro-fluoromethane and dichloro-difluoromethane [10]. Oxygen deficiency in the part of the myocardium may produce high Ca^{2+} and the sufficiently low conductive areas in the ventricles may cause ventricular tachycardia.

The first group of male Wister rats had not received prior treatment, while the second group were subjected to procedures for the induction of pulmonary emphysema. Experimental induction of pulmonary emphysema caused an increase in sensitivity of the rat to cardiopulmonary toxicity from inhalation of trichloro-fluoromethane (FC 11) and dichloro-difluoromethane (FC 12). Emphysematous rats *without anesthesia* were subjected to traumatic stress in the tight cage. The inhalation of 5% FC 11 and 40% FC 12 to the unanesthetized emphysematous rats caused ventricular tachycardia, which was not observed in control unanesthetized nonemphysematous rats during exposure to the same concentration of the fluorocarbons. The acceleration in the heart rate in response to the inhalation of FC 11, or hypoxia in control unanesthetized nonemphysematous rats was prevented by prior treatment with a beta blocking drug, sotalol.

The ventricular tachycardia during inhalation of *unanesthetized* emphysematous rats to FC 11 and FC 12 is thought to be caused by calcium overload, enhanced leak of Ca^{2+} from the SR, inhibition of inwardly rectifying K^+ channels, closings of gap junctional channels by means of traumatic stress, hypoxia, and sympathomimetic toxicity to the fluorocarbons.

10.6 Antiarrhythmic drugs

Drugs may be antiarrhythmic if they can prevent the firing of premature depolarizations in the cardiac muscle cells. The frequency of action potential generation is slowed by raising the threshold, increasing maximum diastolic potential (maximum magnitude of membrane potential in the repolarization phase), or by increasing the action potential duration [11]. DADs can be treated by the agents or conditions which suppress the premature depolarization by the increasing threshold, maximum diastolic potential, or action potential duration as shown in Fig. 10.6.

Parasympathetic stimulation and ACh may activate IKACh channels, which may increase the maximum diastolic potential (see Section 7.3). In addition, an increase in endogenous NO production by activation of muscarinic M_3 receptor on vascular endothelial cells may cause vasodilation and increase oxygen supply to the heart, which is considered to prevent the heart from arrhythmias (see Section 8.7). *β adrenergic stimulation* elevates the cytosolic cyclic AMP level that increases cationic influx called funny currents, I_f (see Section 7.6) and inward Ca^{2+} currents through voltage-gated Ca^{2+} channels, and inhibits inwardly rectifying K^+ currents (see Section 7.2), and enhances the Ca^{2+} leak from the sarcoplasmic reticulum (SR) [5] by modulating RyR2 channels, which causes cardiomyocytes excitable and predispose to tachyarrhythmias.

β blockers inhibit β_1 effects on myocardiocytes. β blockers may lower cytosolic Ca^{2+} level by its inactivating effects on I_f and voltage-gated Ca^{2+} channels, which slows down the rate of heartbeats. β blockades can be antiarrhythmic by preventing premature depolarizations for the following reasons. They avoid ATP-deficiency by reducing heart rate, reduce the cytosolic Ca^{2+} level by lowering cAMP level, remove noises of the signal and prolong its action potential by maintaining the normal activity of inwardly rectifying channel, and protect RyR2 channels from accidental

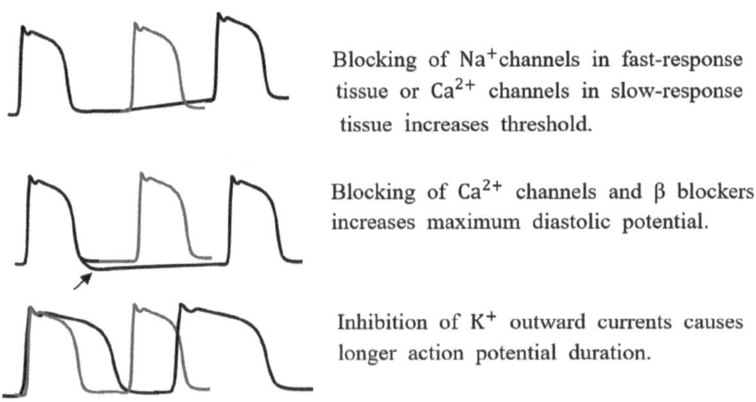

Fig. 10.6 Slowing of the rate of action potential occurs when there is an increased threshold, increased maximum diastolic potential, and increased action potential duration.

opening by blocking sympathetic activity. β blockers are most commonly used to treat DADs. Sotalol is a nonselective β blocker without intrinsic sympathomimetic activity. Since sotalol prolongs the QT interval by reducing outward K^+ currents, it should be avoided to use in conjunction with other medications such as quinidine that prolong the QT interval. Sotalol therapy has been indicated for the management of supraventricular arrhythmias [12]. Since β blockers reduce the frequency of the initiation of action potential in the SA node and reduce the rate of impulse in the AV node when used together with Ca^{2+} channel blockers, they are useful to treat supraventricular tachycardia that involves the AV node.

Lidocaine blocks voltage-gated Na^+ channels. When a premature stimulus is induced in fast-response cells during the repolarizing phase 3, the magnitude of the resultant Na^+ influx depends on the number of Na^+ channels recovered from the inactive state. The relationship curve between the membrane potential and degree of recovery of Na^+ channels from the inactive state can be altered by Na^+ channel blockers such as local anesthetics which causes a longer refractory period. And a greater membrane depolarization is required to bring the partially inactivated Na^+ channels from a resting to an exciting state (Fig. 10.4). Lidocaine has antiarrhythmic effects by blocking voltage-gated sodium channels reversibly, reducing conduction rate in the cardiac muscle fibers, and thus increasing threshold and the longer period of refractoriness. Lidocaine mainly used as local anesthetics can be administered as an acute intravenous therapy of arrhythmias to remove abnormal premature contractions. Lidocaine decreases conduction velocity in fast-response tissue and increases QRS duration by suppressing upstroke velocity when the heart rate is extremely fast.

Quinidine blocks volage-gated Na^+ channels and voltage-gated K^+ channels. The effect of blocking the inward sodium current suppresses the upstroke velocity of the cardiac action potential. It is an open-state blocker of Na^+ channels with a time constant recovery. Quinidine causes a longer refractory period of Na^+ channels and raises the threshold. At higher heart rates, the block increases, while at lower heart rates, the block decreases. Quinidine decreases conduction velocity in fast-response tissue as a consequence of blocking inward Na^+ and outward K^+ currents. At therapeutic concentrations, quinidine also prolongs QT interval up to 25% via blocking outward K^+ currents as side effects (Fig. 10.6). In some myocardial cells and Purkinje fibers, quinidine increases the probability of onset of EADs especially when the serum concentration of K^+ is low. Hence, quinidine may induce Torsades de pointes (polymorphic ventricular tachycardia with increased QT interval) particularly when serum magnesium and potassium are low.

Verapamil blocks voltage-gated Ca^{2+} channels in cardiac cells and its major expected physiological effects reduce the AV node conduction. Since the calcium influx is especially important to generate an action potential in the AV node, the drug usually slows an impulse transmission through the AV node by reducing the generation of action potentials, resulting from the raised threshold. These agents can be used to protect the ventricles from supraventricular tachycardia. β blockers and Ca^{2+} channel blockers are administered together to treat tachycardia.

Rotigaptide has been reported to phosphorylate connexin 43 on gap junctions in the heart [13]. Treatment with rotigaptide probably reduces the occurrence of arrhythmia activating various protein kinase C isoforms which phosphorylates Cx43.

Question 10.1 What causes a decrease in the threshold, and what causes an increase in threshold in the heart?

Answer β_1 adrenergic stimulation by drugs or stress, and mechanical stretch may decrease threshold in both fast- and slow-response cells by elevating the resting membrane potential by increasing the cytosolic Ca^{2+}. Na^+ channel blockers such as lidocaine and quinidine increase the threshold in fast-response cells, and the Ca^{2+} channel blocker such as verapamil elevates the threshold especially in slow-response cells in the heart.

10.7 Discussion

The pathological extra loop circuit may be formed transiently if there is a combination of an area of low conductive tissue as an ectopic focus and high resistive tissues as blocks of ionic currents in the ventricles. The focus in the ventricles initiates a premature depolarization just after the preceding T wave of the pulse from the SA node. The impulse is blocked by the scar tissues and may go around and around to create tachyarrhythmia. The Ca^{2+} overload, valvular heart disease, or cardiomyopathy can predispose ventricular tachyarrhythmias to the ventricular fibrillation.

High Ca^{2+}, low K^+, low PH, ATP-deficiency, myofilament Ca^{2+} sensitization [14], ischemic damage, or anything that increase sympathetic stimulation [15] may predispose latent pacemaker cells in the AV node, the bundle of His, and Purkinje fibers, or cardiac muscle cells to get automaticity. Low conductivity and high Ca^{2+} in the cardiac cells are prerequisites to obtain automaticity. High Ca^{2+} under sympathetic stimulation is considered as a basis of premature contractions since it reduces not only the ion conductivity of the tissue but also makes the muscle excitable. A high concentration of Ca^{2+} in the tissue or a cell under sympathetic stimulation can become a focus of ectopic depolarization. The Ca^{2+} overloaded cells seen in the ischemic or under sympathetic stress becomes more excitable, and the accidental Ca^{2+} discharge from the SR induced by high levels of cytosolic Ca^{2+} [16,17] may initiate premature firings.

During the early stage of hypoxia, the stimulation of the adrenal medulla mediated by the sympathetic nerve [18,19] may sensitize the heart to arrhythmia. Blood vessels distributed in the heart have β_2 receptors. β_2-adrenergic stimulation and adenosine and epinephrine released under hypoxia may elicit beneficial vasodilation and metabolic effects such as stimulation of glycogen breakdown to provide glucose for ATP production via anaerobic glycolysis [20]. And after ischemic preconditioning, connexin 43 is interestingly retained at the intercalated disc and not subjected to subsequent S368 phosphorylation for a closure [21]. The initial stage of hypoxia may contribute somewhat to protect cardiac muscle from a shortage of oxygen. But phosphorylation of RyR2 and ATP-depletion may soon occur via sympathetic stimulation and oxygen deficiency, which may predispose to premature contractions. Administrations of β blocker with enough oxygen supply will reduce consumption of ATP, which then prevents tachyarrhythmias.

10.8 Conclusions

Arrhythmias occur if the disorganized impulse originating in the ectopic focus somewhere in the heart interferes with normal transmission of an action potential from the SA node. The transmission of an impulse from the SA node will be disturbed if ectopic focuses somewhere in and around the atria fire disorganized impulses as seen in atrial fibrillation, or if an ectopic focus fires an impulse during the repolarization phase as seen in Torsades de pointes or just after the preceding T wave of the pulse from the SA node as seen in premature contractions, or if there is an extra anatomically abnormal electric conduction route where a loop pathway happens to be formed by a premature action potential originating in an ectopic focus as seen in supraventricular tachycardias, or if a pathological transient loop circuit around something like scar tissue in the ventricles is formed as seen in ventricular tachyarrhythmia. The Ca^{2+} overload, valvular heart disease, or cardiomyopathy can predispose the ventricular tachyarrhythmia to ventricular fibrillation.

The cell with high Ca^{2+} under sympathetic stimulation or ischemic condition is considered as a candidate for an ectopic focus of the premature contractions since it has not only low ion conductivity but also is excitable.

The generation of premature contractions induced by sympathetic stimulation may be reduced by the administration of an adequate amount of β blockers, and supraventricular tachycardias are controlled by Ca^{2+} channel blockers and β blockers.

References

[1] B. Akdemir, H. Yarmohammadi, M.C. Alraies, W.O. Adkisson, Premature ventricular contractions: reassure or refer? Cleve. Clin. J. Med. 83 (7) (2016) 524–530.
[2] T. Kalogeris, C.P. Baines, M. Krenz, R.J. Korthuis, Cell biology of ischemia/reperfusion injury, Int. Rev. Cell Mol. Biol. 298 (2010) 229–317, https://doi.org/10.1016/B978-0-12-394309-5.00006-7.
[3] S.O. Marx, S. Reiken, Y. Hisamatsu, T. Jayaraman, D. Burkhoff, N. Rosemblit, et al., PKA phosphorylation dissociates FKBP12.6 from the calcium release channel (ryanodine receptor): defective regulation in failing hearts, Cell 101 (2000) 365–376.
[4] D.M. Johnson, G. Antoons, Arrhythmogenic mechanisms in heart failure: linking β-adrenergic stimulation, stretch, and calcium, Front. Physiol. 9 (2018) 1453.
[5] J. Curran, M.J. Hinton, E. Ríos, D.M. Bers, T.R. Shannon, β-adrenergic enhancement of sarcoplasmic reticulum calcium leak in cardiac myocytes is mediated by calcium/calmodulin-dependent protein kinase, Circ. Res. 100 (2007) 391–398.
[6] X. Ai, J.W. Curran, T.R. Shannon, D.M. Bers, S.M. Pogwizd, Ca^{2+}/calmodulin-dependent protein kinase modulates cardiac ryanodine receptor phosphorylation and sarcoplasmic reticulum Ca^{2+} leak in heart failure, Circ. Res. 97 (12) (2005) 1314–1322.
[7] D.A. Gutierrez, M. Fernandez-Tenorio, J. Ogrodnik, E. Niggli, NO-dependent CaMKII activation during β-adrenergic stimulation of cardiac muscle, Cardiovasc. Res. 100 (2013) 392–401.
[8] M.J. Shen, D.P. Zipes, Role of the autonomic nervous system in modulating cardiac arrhythmias, Circ. Res. 114 (2014) 1004–1021.
[9] M. Gheorghiade, K.F. Adams, W.S. Colucci, Digoxin in the management of cardiovascular disorders, Circulation 109 (24) (2004) 2959–2964.
[10] T. Watanabe, D. Aviado, Toxicity of aerosol propellants in the respiratory and circulatory systems: VII. Influence of pulmonary emphysema and anesthesia in the rat, Toxicology 3 (1975) 225–240.
[11] D.M. Roden, Antiarrhythmic drugs, in: The Pharmacological Basis of Therapeutics, eleventh ed., McGraw-Hill, New York, 2006 (Chapter 34).
[12] R. Samanta, A. Thiagalingam, C. Turner, D.J. Lakkireddy, P. Kovoor, The use of intravenous sotalol in cardiac arrhythmias, Heart Lung Circ. 27 (11) (2018) 1318–1326.
[13] L. Axelsen, M. Stahlhut, S. Mohammed, B.D. Larsen, M. Nielsen, N.H. Holstein-Rathlou, S. Andersen, O. Jensen, J. Hennan, A. Kjølbye, Identification of ischemia-regulated phosphorylation sites in connexin43: a possible target for the antiarrhythmic peptide analogue rotigaptide (ZP123), J. Mol. Cell. Cardiol. 40 (6) (2006) 790–798.
[14] S. Huke, R. Venkataraman, M. Faggioni, S. Bennuri, H.S. Hwang, F. Baudenbacher, B.C. Knollmann, Focal energy deprivation underlies arrhythmia susceptibility in mice with calcium-sensitized myofilaments, Circ. Res. 112 (10) (2013) 1334–1344.

[15] A.L. Wit, B.F. Hoffman, P.F. Cranefield, Slow conduction and reentry in the ventricular conducting system: I. Return extrasystole in canine Purkinje fibers, Circ. Res. 30 (1972) 1–10.
[16] R. Zucchi, S. Ronca-Testoni, The sarcoplasmic reticulum Ca^{2+} channel/ryanodine receptor: modulation by endogenous effectors, drugs and disease states, Pharmacol. Rev. 49 (1) (1997) 1–52.
[17] L.A. Venetucci, A.W. Trafford, S.C. O'Neill, D.A. Eisner, The sarcoplasmic reticulum and arrhythmogenic calcium release, Cardiovasc. Res. 77 (2008) 285–292.
[18] T. Watanabe, Diffusion and flow, and respiratory system, in: Biophysical Basis of Physiology and Calcium Signaling Mechanism in Cardiac and Smooth Muscle, Academic Press, San Diego, CA, 2018 (Chapter 5).
[19] R.S. Brody, T. Watanabe, D.M. Aviado, Toxicity of aerosol propellants on the respiratory and circulatory systems, Toxicology 2 (1974) 173–184.
[20] A. Gorlach, Control of adenosine transport by hypoxia, Circ. Res. 97 (2005) 1–3.
[21] J.L. Solan, P.D. Lampe, Connexin 43 phosphorylation-structural changes and biological effects, Biochem. J. 419 (2) (2009) 261–272.

Reference data

Fundamental Constants

Speed of light in vacuum	c	3.00×10^{8} m s^{-1}
Avogadro's number	N_A	6.02×10^{23} mol^{-1}
Gas constant	R	8.314 J K^{-1} mol^{-1}
Boltzmann's constant	k	1.38×10^{-23}J K^{-1}
Charge on electron	e	1.60×10^{-19} C
Planck's constant	h	6.63×10^{-34} J s
Faraday's constant	F	9.65×10^{4} C mol^{-1}
Electron rest mass	m_e	9.11×10^{-31} kg
Proton rest mass	m_p	1.6726×10^{-27} kg
Neutron rest mass	m_n	1.6749×10^{-27} kg
Atomic mass unit (1 U)	m_n	1.6605×10^{-27} kg

Electronegativity of Some Elements

	H 2.20																He	
2	Li 0.98	Be 1.57											B 2.04	C 2.55	N 3.04	O 3.44	F 3.98	Ne
3	Na 0.93	Mg 1.31											Al 1.61	Si 1.90	P 2.19	S 2.58	Cl 3.16	Ar
4	K 0.82	Ca 1.00	Sc 1.36	Ti 1.54	V 1.63	Cr 1.66	Mn 1.55	Fe 1.83	Co 1.88	Ni 1.91	Cu 1.90	Zn 1.65	Ga 1.81	Ge 2.01	As 2.18	Se 2.55	Br 2.96	Kr 3.00
5	Rb 0.82	Sr 0.95	Y 1.22	Zr 1.33	Nb 1.6	Mo 2.16	Tc 1.9	Ru 2.2	Rh 2.28	Pd 2.20	Ag 1.93	Cd 1.69	In 1.78	Sn 1.96	Sb 2.05	Te 2.1	I 2.66	Xe 2.60
6	Cs 0.79	Ba 0.89	*	Hf 1.3	Ta 1.5	W 2.36	Re 1.9	Os 2.2	Ir 2.20	Pt 2.28	Au 2.54	Hg 2.00	Tl 1.62	Pb 1.87	Bi 2.02	Po 2.0	At 2.2	Rn 2.2

Index

Note: Page numbers followed by *f* indicate figures and *t* indicate tables.

F

G

H

I

L

M

N

FSC
www.fsc.org
MIX
Papier | Fördert gute Waldnutzung
FSC® C083411

Kolibri 360 GmbH
Ferdinand-Jühlke-Straße 7
99095 Erfurt, Deutschland
produktsicherheit@kolibri360.de